Ozone in the
Free Atmosphere

Ozone in the Free Atmosphere

Edited by

Robert C. Whitten

NASA-Ames Research Center
Moffett Field, California

Sheo S. Prasad

Jet Propulsion Laboratory
Pasadena, California
and
University of Southern California

VNR VAN NOSTRAND REINHOLD COMPANY
————————————————————— *New York*

The editors dedicate this book to their late colleague, Louis A. Capone.

Copyright © 1985 by **Van Nostrand Reinhold Company Inc.**
Library of Congress Catalog Card Number: 84-17254
ISBN: 0-442-29207-4

Manufactured in the United States of America.

Published by Van Nostrand Reinhold Company Inc.
135 West 50th Street
New York, New York 10020

Van Nostrand Reinhold Company Limited
Molly Millars Lane
Wokingham, Berkshire RG11 2PY, England

Van Nostrand Reinhold
480 Latrobe Street
Melbourne, Victoria 3000, Australia

Macmillan of Canada
Division of Gage Publishing Limited
164 Commander Boulevard
Agincourt, Ontario MIS 3C7, Canada

15 14 13 12 11 10 9 8 7 6 5 4 3 2 1

Library of Congress Cataloging in Publication Data
Main entry under title:
Ozone in the free atmosphere.
   Includes bibliographies and index.
   1. Atmospheric ozone.   2. Stratosphere.   3. Troposphere.
I. Whitten, R. C. (Robert Craig), 1926- .
II. Prasad, Sheo S.
QC879.7.094   1985   551.5'112   84-17254
ISBN 0-442-29207-4

Contents

Preface

Ozone, although it is a minor species in the atmosphere, is of considerable importance to mankind by virtue of its acting as a shield over the biosphere against lethal ultraviolet radiation from the sun. In addition, ozone may have significant links to weather and climate. Events of the past decade have generated a widespread awareness of the importance of ozone from an altogether new perspective, namely, the possibility that human activities may pollute the stratosphere and thereby significantly disturb the ozone layer. In 1971 concern was expressed that the operation of aircraft in the stratosphere could, due to the release of nitrogen oxides, lead to a serious reduction of high altitude ozone and thus an increase of harmful solar ultraviolet radiation intensity at the earth's surface. The result was the establishment of the Climatic Impact Assessment Program (CIAP) within the United States and corresponding programs in other nations for the purpose of assessing the threat. Our knowledge of ozone, its geographical and temporal variations, its photochemistry, and its transport increased rapidly during that era. Although CIAP ended with the publication of its findings in 1975, other agencies and programs (for example, the High Altitude Pollution Program [HAPP] of the Federal Aviation Administration, various programs of the National Aeronautics and Space Administration, and the World Meteorological Organization, to name a few) have continued to support research on atmospheric ozone. Such continuation was necessitated partly because CIAP left numerous important questions on ozone photochemistry, transport, and so on unresolved and partly because of the realization that free chlorine released in the stratosphere by photodissociation of chlorofluoromethanes could also seriously deplete stratospheric ozone. It is interesting that as our knowledge of ozone photochemistry advanced, the assessment of the seriousness of the potential due to the release of nitrogen oxides in the stratosphere at first decreased steadily, and the seriousness of the chlorine problem seemed to increase. But in 1981 certain new developments in the chemical kinetics of ozone reversed the situation. Nitrogen oxide release is now deemed to be the more serious threat under certain conditions. Of course, the situation may again change in a similarly radical manner. An interesting account of the research carried out on the threats to ozone by nitrogen oxide and chlorine release has been presented by Dotto and Schiff in *The Ozone War* (Doubleday, 1978).

The present book has a different purpose. It summarizes the state of scientific knowledge of stratospheric and free tropospheric ozone as it exists at the beginning of 1983. Chapter 2 is on ozone measurement and geographical and temporal variations and discusses measurement techniques; it presents and evaluates the data on ozone distributions including the recent satellite observations. Chapters 2 and 3 treat respectively the photochemistry and the transport of stratospheric ozone; they describe ozone formation and distribution from the standpoint of the most recent major advances. Chapter 4 treats ozone in the free troposphere from the standpoint of photochemistry and transport, while Chapter 5 describes current assessments of possible threats to the ozone layer. The text concludes with a discussion of the effects on health and climate of decreasing stratospheric ozone abundance. It is important, however, to point out that it is beyond the scope of the book to discuss the influence of changes in stratospheric ozone on atmospheric circulation. This is a major subject in its own right and is more appropriately discussed in a text on the general circulation of the atmosphere.

The editors are indebted to the authors of the various chapters for close cooperation and timely submission of their manuscripts.

ROBERT C. WHITTEN
SHEO S. PRASAD

Contributors

Dr. Edwin F. Danielsen
Mail Stop 245-3, NASA-Ames Research Center, Moffett Field, California 94035

Dr. Jack Fishman
Mail Stop 401B, NASA-Langley Research Center, Hampton, Virginia 23665

Professor Julius London
Department of Geoastrophysics, University of Colorado, Boulder, Colorado 80309

Dr. Fred Luther
Lawrence Livermore Laboratory, 7000 East Avenue, Livermore, California 94550

Dr. Sheo S. Prasad
Jet Propulsion Laboratory, 4800 Oak Grove Avenue, Pasadena, California 91109

Dr. Richard P. Turco
R and D Associates, 4640 Admiralty Way, Marina del Rey, California 90291

Dr. Robert C. Whitten
Mail Stop 245-3, NASA-Ames Research Center, Moffett Field, California 94035

Ozone in the
Free Atmosphere

Introduction

Robert C. Whitten
NASA-Ames Research Center

Sheo S. Prasad
Jet Propulsion Laboratory
California Institute of Technology
and
Department of Physics
University of Southern California

Ozone, which shields the biosphere from the lethal ultraviolet radiation from the sun, is probably the youngest member of the family of gases that comprise the present natural atmosphere of the earth. (Gases like the chlorofluoromethanes entered our atmosphere only yesterday [speaking on geological time scales]; however, they did so not by any natural process but by anthropogenic activities.) More than 4.5 billion years ago, the sun, the earth, and the rest of the solar system condensed out of an interstellar cloud of gas and dust. As the heavier elements (such as silicon, iron, nickel) of the primordial solar nebula coalesced to form the solid earth, the lighter elements (carbon, nitrogen, oxygen) formed simple chemical compounds (methane, ammonia, water) with hydrogen, which was overwhelmingly the most abundant element in the nebula. These volatile compounds of hydrogen, carbon, nitrogen, and oxygen were physically and chemically trapped within the solid earth. Early in the history of the solar system, the primordial atmosphere of the earth (i.e., the remnants of the gaseous nebula gravitationally attached to the earth) was swept away by the intense solar wind during the *T Tauri* phase of the sun (Cameron, 1973; Hayashi, 1961). In the course of time, however, a new atmosphere formed around the earth whose ingredients were the trapped volatile compounds released or "outgassed" from the solid earth (water, carbon dioxide, chlorine, nitrogen, sulfur, etc.; see Rubey, 1955 for details).

Condensation of the outgassed water vapor led to the formation of the oceans. The bulk of the outgassed carbon dioxide then readily entered the oceans in a dissolved state to form carbonate and bicarbonate ions (see, for example, Walker, 1977). Thus, only a trace of CO_2 remained in the atmosphere. Chlorine and sulfur also met the same fate. Hence, the present atmosphere of the earth is secondary in origin (see, however, Urey [1952a, 1952b] for an altogether different scenario of the origin of the earth's ancient atmosphere).

It is important to note that oxygen (O_2), the second most abundant constituent in the present atmosphere, was not present in the primitive atmosphere, because it was not outgassed from the earth's interior (Rubey, 1955). According to one school of thought (Walker, 1978; Kasting, Liu, and Donahue, 1979; Vander Wood and Thiemens, 1980) emergence of oxygen had to wait for about 2 billion years, until the advent of green plant photosynthesis, because the photodissociation of the atmospheric water vapor in the prebiological atmosphere would have produced O_2 of order 10^{-12} PAL (present atmospheric level) or less, unless the volcanic release of hydrogen was much less than at present. In spite of these uncertainties, it is clear that oxygen reached its present level only in very recent times. For instance, at the beginning of the Cambrian Era, some 580 million years ago, atmospheric O_2 abundance was only 0.2 PAL (Rhoads and Morse, 1971). Ozone appeared in the atmosphere only after the appearance of molecular oxygen. As the atmospheric levels of O_2 increased, so did the levels of ozone (O_3). It has been argued (Berkner and Marshall, 1965) that the evolution of ozone controlled the migration of life from the safety of oceans onto the land. According to Berkner and Marshall's (1965) qualitative calculations, the "first critical level" of O_3 was reached when O_2 attained a level of 10^{-2} PAL. At this stage the O_3 column density ($\sim 1 \times 10^{18}$ molecules cm^{-2}) was sufficient to restrict the ultraviolet zone of lethality to a thin layer close to the ocean surface. This greatly enhanced photosynthetic activity by permitting life to spread up to the vicinity of the ocean surface. The second critical level of O_3 was reached when O_2 was at 10^{-1} PAL. An ozone column density of about 5.6×10^{18} cm^{-2} at this stage was sufficient to absorb totally the lethal ultraviolet radiation and to enable life to migrate onto the land for the first time. More recent calculations (e.g., Levine, 1982) give approximately the same relationship between the ozone column content and the O_2 level. Nevertheless, Margulis, Walker, and Rambler (1976) and Rambler and Margulis (1980) have questioned Berkner and Marshall's scenario. In agreement with Walker (1977) and many others we also wish to emphasize that very little is known about the geological history of the earth's atmosphere. Nevertheless, this much is probably certain: that ozone is perhaps the youngest member of the natural atmosphere and that the emergence of the ozone layer was instrumental in the emergence of higher forms of life on the land.

CHEMICAL AND ATMOSPHERIC DISCOVERY OF OZONE

Although molecular nitrogen and oxygen were discovered as early as 1774 (see Weeks, 1968 for an historical account), the recognition of ozone as a distinct chemical species came only after the advent of controlled electric energy. Apparently, the Dutch scientist Van Marum was the first to observe that a peculiar odor resulted from passing an electrical discharge through oxygen. The substance causing the odor was identified in the laboratory by the Swiss chemist Schönbein, who noted that the same strong odor occurred in oxygen generated from the electrolysis of water and in air when they were subjected to an electric discharge. He suggested that the substance might be a permanent feature of the atmosphere and thus deserved a name; he (Schönbein, 1840) proposed that it be called ozone (probably after the Greek word *ozien* = to smell). More precise knowledge of the origin of ozone was obtained a few years later when de la Rive and Marignae (de la Rive, 1845) demonstrated its production by electric discharge in pure and dry oxygen. The first chemical identification of ozone was probably due to Soret (1863) who stated that "la molécule d'ozone fût composée de 3 atomes OOO et constituât un bioxyde d'oxygène."

Existence of ozone in the natural troposphere was chemically proven by Houzeau (1858) with the aid of a test paper containing potassium iodide, which was developed by Schönbein. This observation was soon confirmed by many other investigators, including the famed American oceanographer Matthew Fontaine Maury who suggested that ozone might serve as tracer of air currents. The first clear spectroscopic detection of ozone related to the atmosphere was made in 1880 by Chappuis and Hautefeuille (Chappuis, 1880) who were able to measure eleven absorption bands attributable to this substance and coinciding with the telluric bands in the solar spectrum that were thought to be due to the selective absorption properties of the earth's atmosphere. During the same period Hartley (1881a) detected in the laboratory the strong ultraviolet spectrum of ozone below 300 nm (which now bears his name). Hartley (1881b) also demonstrated that the atmospheric limit of the solar ultraviolet spectrum detected by Cornu (1879a, 1879b) was due to ozone. Later, Fowler and Strutt (Lord Rayleigh) (1917) proved that the bands discovered by Huggins and Huggins (1890) in the spectrum of

Sirius were caused by the absorption in the earth's atmosphere and that the absorber was ozone.

Because ozone was known to be produced efficiently by electric discharge, including lightning, the early belief was that ozone is distributed close to the earth's surface. Hartley (1881b) was the first to point out that ozone is a normal constituent of the higher atmosphere and that it is in larger proportion there than near the earth's surface. The first satisfactory determination of the height of the absorbing medium was given by Lord Rayleigh (Strutt, 1918), who based his estimate on observations of the solar spectrum at sunrise and sunset. He concluded that atmospheric ozone is largely confined to a layer between 40 and 60 km above sea level. Because of limitations of the experimental method, his estimate was only qualitatively correct. Later, greatly improved measurements by Götz, Meetham, and Dobson (1934) established the presently accepted bounds to the stratospheric ozone layer, 15 km to 50 km altitude with a maximum occurring near 25 km. The Regeners (1934) confirmed the Götz, Meetham, and Dobson measurements by making direct spectrographic observations from a balloon that ascended to 31 km. The measured solar ultraviolet spectrum showed clearly that ascent through the ozone layer resulted in gradual extension of the spectrum toward the ultraviolet. This is not to say that ozone is absent from the lower atmosphere. It is an important minor constituent of the troposphere, a point discussed at length in Chapter 4.

The first technique for measuring the total abundance of ozone in a vertical column was suggested by Fabry and Buisson (1921). Dobson (Dobson and Harrison, 1926; Harrison and Dobson, 1925) refined the method and developed instrumentation capable of making precise measurements; their technique is even today the standard for making ground-based measurements of total ozone. Basically, the Dobson instrument measures relative intensities at different wavelengths in the spectral range 300-340 nm after passage of the radiation through the atmosphere. From the data so obtained, the ozone column density is "unfolded" (see Chapter 1 for further discussion). Ozone observations are also made with a broad band optical filter, mainly in the Soviet Union and Eastern Europe (Gushchin, 1974). However, those data are not always in agreement with Dobson data taken at the same place and do not seem to be as precise as the latter. For these reasons, only the Dobson instrument at Boulder, Colorado has been accepted as the international standard.

Techniques have also been developed to measure ozone in situ from aircraft and balloon platforms. These methods use a "wet" chemical system in which ozone reacts with a chemical (potassium iodide) in solution, or a gas phase system in which the reactions produce chemiluminescence of which the intensity can be measured. The most recent developments employ satellites that either measure the spectral radiance of backscattered solar ultraviolet (BUV) radiation or limb scan in which the absorption of solar radiation over an appropriate spectral segment is measured. These approaches will be discussed at length in Chapter 1.

Soon after the discovery of the ozone layer in the upper atmosphere, the question of its formation arose. Warburg (1921) found that light of wavelength 253 nm is able to dissociate O_2, leading to the formation of two oxygen atoms. Actually, 253 nm is close to the limit of the O_2, dissociation spectrum. Only much later did Herzberg (1952) discover that the solar spectrum (195-260 nm) may penetrate deeply into the atmosphere and dissociate molecular oxygen in the O_2 bands and continuum that bear his name. Nevertheless, it was known by 1930 that solar radiation could penetrate to the altitudes of the ozone layer, and rough estimates of the corresponding O_2 photolysis cross sections were available. This knowledge led Chapman (1930) to propose a series of processes, subsequently referred to as the Chapman mechanism, that could explain the formation of the ozone layer, Chapman's predictions were in excellent agreement with observations available at the time.

The Chapman reactions are

$$O_2 + h\nu \rightarrow O + O \tag{1}$$
$$O + O_2 + M \rightarrow O_3 + M \tag{2}$$
$$O_3 + h\nu \rightarrow O + O_2 \tag{3}$$
$$O + O_3 \rightarrow O_2 + O_2 \tag{4}$$

in which M is a third body required to carry off the excess energy of the association process. The rapid cycle composed of reactions (2) and (3) does not, of course, destroy ozone; it merely transfers an oxygen atom between a free state and a state in which it is bound to an oxygen molecule. Reaction (4) is the only ozone loss process. At the time of Chapman's proposal, the rate coefficient of reaction (4) was not known and an estimate had to be used. Recent measurements (e.g., McCrumb and Kaufman, 1972) have established that reaction (4) is insufficient to account for all the ozone destruction necessary to reconcile theory with observation. Many years after Chapman's original work, it was suggested that catalytic processes can accelerate reaction (4)

$$X + O_3 \rightarrow XO + O_2 \tag{5}$$
$$XO + O \rightarrow X + O_2 \tag{6}$$

$$\overline{O + O_3 \rightarrow 2\,O_2} \quad \text{(overall reaction)}$$

Note that chemical agent X is not consumed in the process and can effectively destroy many ozone molecules. Suggested candidates for X are OH (Hampson, 1964; Hunt, 1966), NO (Crutzen, 1970), and Cl (Stolarski and Cicerone, 1974). While all three catalysts are important, NO is the dominant ozone-active catalyst in the unperturbed stratosphere (e.g., Turco et al., 1978). A detailed discussion of ozone photochemical processes will be given in Chapter 2.

So far, only the photochemistry of ozone has been discussed. Actually, ozone can be transported vertically and horizontally by atmospheric motions from regions of formation to regions where it can be stored; for example, from above ~30 km to lower altitudes where photochemical destruction mechanisms are weak. Ozone is transported horizontally in a rather complex manner that is dependent upon meteorological conditions and on geography, especially latitude. Detailed discussion of ozone transport is the subject of Chapter 3.

Because of the apparent effectiveness of catalytic agents in depleting ozone, scientists have been concerned about anthropogenic threats to the ozone layer. Johnston (1971) suggested that nitrogen oxides deposited in the upper atmosphere by supersonic transport aircraft could degrade ozone abundance to unacceptable levels. In response to this threat, the U.S. government established and funded the Climatic Impact Assessment Program (CIAP) under the U.S. Department of Transportation (Grobecker, Coroniti, and Cannon, 1974). This effort was the first large-scale study of stratospheric ozone problems.

Around this time, Molina and Rowland (1974) suggested that free chlorine introduced into the upper atmosphere as a result of the photodissociation of man-made chlorofluoromethanes (CFMs) could also greatly enhance the catalytic destruction of ozone. CFMs were in wide use as refrigerants and aerosol propellants, and researchers became concerned that they might accumulate to high levels in the atmosphere. These and other threats to ozone are discussed in Chapter 5.

IMPORTANCE OF OZONE IN THE UPPER ATMOSPHERE

Perhaps the most important aspect of the ozone layer, at least to the biosphere, is its absorption of solar ultraviolet radiation before it can reach the earth's surface. It has been known for a long time that ultraviolet radiation can be harmful to both animal and plant life. Its most striking effect is sunburn or erythema in humans. Repeated exposure to ultraviolet radiation may result in the occurrence of skin cancer, which can be broadly divided into (potentially lethal) melanoma and nonmelanoma. There is strong evidence (see Fears, Scotto, and Schneiderman, 1974) that the occurrences of both types of cancer are latitude-dependent and that they correlate well with the mean ozone overburden at each latitude. The connection between stratospheric ozone abundance and the effects on the biosphere will be developed in Chapter 6.

Because ozone is a strong absorber of solar radiation, its presence in the upper atmosphere also results in local heating there. Indeed, it is the presence of ozone at high altitudes that leads to the formation of the *stratosphere*. The rate of energy deposition per unit mass in the ozone reaches a maximum at about 45 to 50 km altitude (near the *stratopause*). A minimum in temperature (the *tropopause*) occurs at lower altitudes (8 to 16 km, depending upon latitude) so that the intervening layer is characterized by a strong temperature inversion and dynamically stable air. A second temperature minimum occurs at about 85 km altitude (the *mesopause*). The regions that intervene between the surface, the tropopause, the stratopause, and the mesopause have been named the *troposphere*, the *stratosphere*, and the *mesosphere* respectively. Atmospheric thermal structure is illustrated in Figure 1.

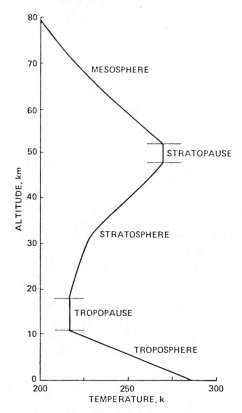

Figure 1. Thermal structure of the atmosphere below 80 km altitude.

Although this book is concerned principally with stratospheric ozone, the occurrence, chemistry, and distribution of ozone in the free troposphere is also of considerable interest both as a tracer of air motions and as a photochemical agent and an air pollutant. It is known, for example, that while part of the tropospheric ozone is produced by photochemical ("smog") reactions in the lower atmosphere, a second significant source is downward transport from the stratosphere. Tropospheric ozone photochemistry will be discussed at length in Chapter 4, and downward transport of ozone from the stratosphere will be treated in Chapter 3.

REFERENCES

Berkner, L. V., and L. C. Marshall, 1965, On the origin and rise of oxygen concentration in the Earth's atmosphere, *J. Atmos. Sci.* **22**:225-261.

Cameron, A. G. W., 1973, Accumulation processes in the primitive solar nebula, *Icarus* **18**:407-450.

Chapman, S., 1930, A theory of upper atmospheric ozone, *R. Meteorol. Soc. Mem.* **3**:103-125.

Chappuis, M. J., 1880, Sur le spectre d'absorption de l'ozone, *Acad. Sci. (Paris) C.R.* **91**:985-986.

Cornu, M. A., 1879*a*, Sur la limite ultra-violette du spectre solaire, *Acad. Sci. (Paris) C.R.* **88**:1101-1108.

Cornu, M. A., 1879*b*, Sur l'absorption par l'atmosphères des radiations ultraviolettes, *Acad. Sci. (Paris) C.R.* **88**:1285-1290.

Crutzen, P. J., 1970, The influence of nitrogen oxides on the atmospheric ozone content, *R. Meteorol. Soc. Q. J.* **96**:320-325.

de la Rive, G., 1845, Sur l'ozone, *Acad. Sci. (Paris) C.R.* **20**:1291.

Dobson, G. M. B., and D. N. Harrison, 1926, Measurement of the amount of ozone in the Earth's upper atmosphere and its relation to other geophysical conditions, *R. Soc. (London) Proc. ser. A,* **110**:660-693.

Fabry, C., and M. Buisson, 1921, Étude de l'extrémité ultraviolette du spectre solaire, *J. Phys. Rad.* **2**:197-226.

Fears, T. J., J. Scotto, and M. Schneiderman, 1974, The Epidemiology of Skin Cancer, paper presented at the 45th Aerospace Medical Association Meeting, Washington, D.C.

Fowler, A., and R. J. Strutt, 1917, Absorption bands of atmospheric ozone in the spectra of sun and stars, *R. Soc. (London) Proc., ser. A,* **93**:577-586.

Götz, F. W. P., A. R. Meetham, and G. M. B. Dobson, 1934, The vertical distribution of ozone in the atmosphere, *R. Soc. (London) Proc., ser. A,* **145**: 416-446.

Grobecker, A. J., S. C. Coroniti, and R. H. Cannon, 1974, *Report of Findings: The Effects of Stratospheric Pollution by Aircraft,* Report DOT-TSC-75-50, U.S. Department of Transportation, Washington, D.C.

Gushchin, G. P., 1974, *Comparison of Ozonometric Instruments in Actinometry,*

Atmospheric Optics, Ozonometry, Trudy GGO no. 279, Israel Program for Scientific Translations, Jerusalem.

Hampson, J., 1964, *Photochemical Behavior of the Ozone Layer,* TN 1627/64, Canadian Armament Research and Development Establishment, Valcartier, Quebec, Canada.

Harrison, D. N., and G. M. B. Dobson, 1925, Measurements of the amount of ozone in the upper atmosphere, *R. Meteorol. Soc. Q. J.* **51**:363.

Hartley, W. N., 1881*a*, On the absorption of solar rays by atmospheric ozone, Part I, *Chem. Soc. (London) J.* **39**:111-119.

Hartley, W. N., 1881*b*, On the absorption of solar rays by atmospheric ozone, Part II, *Chem. Soc. (London) J.* **39**:119-128.

Hayashi, C., 1961, Stellar evolution in early phases of gravitational contraction, *Astron. Soc. Japan Publ.* **13**:450-452.

Herzberg, G., 1952, Forbidden transitions in diatomic molecules II. The $^3\Sigma_u^+ \leftarrow {}^3\Sigma_g^+$ absorption bands of the oxygen molecule, *Can. J. Phys.* **30**: 185-210.

Houzeau, M. A., 1858, Preuves de la présence dans l'atmosphère d'un nouveau principe gazeux, l'oxygène naissant, *Acad. Sci. (Paris) C.R.* **46**:89.

Huggins, W., and Mrs. Huggins, 1890, On a new group of lines in the photographic spectrum of Sirius, *R. Soc. (London) Proc.* **48**:216.

Hunt, B. G., 1966, Photochemistry of ozone in a moist atmosphere, *J. Geophys. Res.* **71**:1385-1398.

Johnston, H. S., 1971, Reduction of stratospheric ozone by nitrogen oxide catalysts from supersonic transport exhaust, *Science* **173**:517-522.

Kasting, J. F., S. C. Liu, and T. M. Donahue, 1979, Oxygen levels in the prebiological atmosphere, *J. Geophys. Res.* **84**:3097-3107.

Levine, J. S., 1982, The photochemistry of the paleoatmosphere, *J. Mol. Evol.* **18**:161-172.

McCrumb, J. L., and F. Kaufman, 1972, Kinetics of the $O + O_3$ reaction, *J. Chem. Phys.* **57**:1270-1276.

Margulis, L., J. C. G. Walker, and M. Rambler, 1976, Reassessment of roles of oxygen and ultraviolet light in pre-cambrian evolution, *Nature* **264**:620-624.

Molina, M. J., and F. S. Rowland, 1974, Stratospheric sink for chlorofluoromethanes: chlorine atom-catalyzed destruction of ozone, *Nature* **249**:810-812.

Rambler, M. B., and L. Margulis, 1980, Bacterial response to ultraviolet irradiation under anearobiosis: Implications for pre-phanerozoic evolution, *Science* **210**:638-640.

Regener, E., and V. H. Regener, 1934, Aufnahmen des ultravioletten Sonnenspectrums in der Stratosphäre und die vertikale Ozonverteilung, *Phys. Z.* **35**: 788-793.

Rhoads, D. C., and J. W. Morse, 1971, Evolutionary and ecological significance of oxygen-deficient marine basin, *Lethaia* **4**:413-428.

Rubey, W. W., 1955, Development of the hydrosphere and atmosphere, with special composition of the early atmosphere, in *Crust of the Earth,* Geological Society of America, Special Paper 62, Boulder, Colo., pp. 631-651.

Schonbein, M., 1840, Recherches sur la nature de l'odeur aui se manifeste dans certaines actions chimiques, *Crit. Rev.* **10**:706.

Soret, M. J.-L., 1863, Sur les relations volumetriques de l'ozone, *Acad. Sci. (Paris) C.R.* **57**:604-609.

Stolarski, R. S., and R. J. Cicerone, 1974, Stratospheric chlorine: a possible sink for ozone, *Can. J. Chem.* **52**:1610-1615.

Strutt, R. J., 1918, Ultraviolet transparency of the lower atmosphere and its relative poverty in ozone, *R. Soc. (London) Proc., ser. A,* **94**:260-268.

Turco, R. P., R. C. Whitten, I. G. Poppoff, and L. A. Capone, 1978, SST's, nitrogen fertilizer and stratospheric ozone, *Nature* **276**:805-807.

Urey, H. C., 1952a, *The Planets: Their Origin and Development,* Yale University Press, New Haven, Conn., p. 245.

Urey, H. C., 1952b, On the early chemical history of the Earth and the origin of life, *Natl. Acad. Sci. (USA) Proc.* **38**:351-352.

Vander Wood, T. B., and M. H. Thiemens, 1980, The fate of the hydroxyl radical in the Earth's primitive atmosphere and implications for the production of molecular oxygen, *J. Geophys. Res.* **85**:1605-1610.

Walker, J. C. G., 1977, *Evolution of the Atmosphere,* Macmillan, New York, p. 318ff.

Walker, J. C. G., 1978, Oxygen and hydrogen in the primitive atmosphere, *Pure and Appl. Geophys.* **116**:222-231.

Warburg, E., 1921, Energy relations in photochemical processes in gases, *Z. Electrochem.* **27**:133-142.

Weeks, M. E., 1968, *Discovery of the Elements,* 7th ed., published by the *Journal of Chemical Education,* Easton, Pa., p. 177.

Chapter 1

The Observed Distribution of Atmospheric Ozone and Its Variations

Julius London

Department of Astro-Geophysics
University of Colorado

Ozone was first noted in air at the end of the eighteenth century as an electrical odor; it was clearly identified as an atmospheric constituent by Schönbein (1840). However, only occasional attempts were made to measure ozone routinely as a regular component of air until early in the twentieth century, when C. G. Abbott and his colleagues at the Smithsonian Institution became involved in a program of measurements of the atmospheric transmittance of solar radiation in the Chappuis ozone bands as part of their solar constant measurement program (see *Annals of the Astrophysical Observatory,* vol. 3, 1914; vol. 4, 1922).

The earliest measurements of atmospheric ozone were made by exposing paper that had been impregnated with potassium iodide in a solution containing starch. A qualitative scale was developed by Schönbein (Khrgian, 1975) such that the blueness of the exposed paper indicated the relative quantity of ozone in the air sample on a scale of 1 to 10. After the discovery by Cornu (1879) of the ultraviolet cutoff of stellar radiation and the laboratory determination of the ozone absorption in the spectral region below 300 nm by Hartley (1880a, 1881a) and in the visible by Chappuis (1880, 1882), it was possible to measure the integrated amount of ozone in a vertical column from the ground to the top of the atmosphere (e.g., Fabry and Buisson, 1913; Cabannes and Dufay, 1927). A routine observing program was first set up by G. M. B. Dobson and co-workers at

Oxford in 1924 and later extended to a number of stations in Europe and Australia (Dobson, Harrison, and Lawrence, 1927, 1929). Analysis of the observations from these stations quickly established the general pattern of seasonal, latitudinal, and meteorological variations in total ozone.

The early observations suggested that the maximum ozone concentration was at a height of about 50 km. However, by a method described below, it was shown from measurements made in 1929 in Spitzbergen that the height of maximum ozone was close to about 22-25 km (Götz, 1931). This measurement was later confirmed by direct spectrographic observations of the solar spectrum made from stratospheric balloon platforms (Regener and Regener, 1934). The major drive for the present global ozone observing system came from the program for the International Geophysical Year (IGY) 1957-1958, when there were approximately 45 stations taking almost daily total ozone measurements. Both the total amount and the vertical distribution of atmospheric ozone are presently measured routinely by various ground-based and satellite-borne instruments and by instruments mounted on different carrier platforms such as balloons, aircraft, and rockets. These measurements are made almost continuously, as in the case of some satellites, or almost daily, as with many of the ground-based observations. They have broad geographic coverage and provide information on the ozone concentration from the ground up to about 90 km. The basic methods currently used for ozone measurements involve optical or chemical techniques. (For a detailed history of early ozone observations see Khrgian, 1975; Nicolet, 1975; London and Angell, 1982, and the historical bibliography at the end of this chapter.)

TECHNIQUES OF OBSERVING ATMOSPHERIC OZONE

It is useful to define the various quantities used in ozone observations and in describing ozone distributions.

The ozone density can be given as the mass density ρ_3 (g cm^{-3}) [or $\gamma(\mu$g m^{-3}) for near surface observations] or the number density n_3 (molecules cm^{-3}). The total number of molecules in a vertical column from height z to the top of the atmosphere is

$$N_3 = \int_z^\infty n_3 \, dz \qquad (\text{cm}^{-2})$$

or, if reduced to standard temperature and pressure,

$$X_3 \quad (\text{STP}) = \frac{N_3}{L_0}$$

where L_0 is Loschmidt's number $= 2.69 \times 10^{19}$ molecules cm^{-3}. For a column of air that contains approximately 8×10^{18} ozone molecules cm^{-2} (the global average total amount of ozone per cm^2 above sea level) the equivalent depth for total ozone is $X_3 \approx 0.3$ cm (STP) or 300 m-atm cm, defined as 300 Dobson units (1 DU $= 10^{-3}$ cm of ozone at STP).

The partial pressure of ozone is p_3 (nb). The mass mixing ratio is $r_3 = \rho_3/\rho$ where ρ is the air mass density; r_3 is usually given as μg g^{-1} (ppmm). The number mixing ratio is $r_3' = n_3/n$ (ppmv) where n is the molecular density of air. Note that $r_3' = p_3/p \approx 0.6r_3$.

When ozone is measured spectroscopically, use is made of Beer's Law, assuming that the coefficients are not path dependent. Then for pure absorption, we have

$$L_\lambda = L_{0\lambda}e^{-\int_0^s k_\lambda \rho_3 \, ds} = L_{0\lambda}e^{-\sigma_\lambda N_3^*} = L_{0\lambda}10^{-a\lambda X_3^*}, \qquad (1\text{-}1)$$

where L_λ and $L_{0\lambda}$ are the observed radiance and the source radiance respectively at wavelength λ; k_λ is the Naperian mass absorption coefficient (cm^2 g^{-1}); σ_λ is the molecular absorption cross section (cm^2), and a_λ is the decadic absorption coefficient (cm^{-1}) derived from laboratory absorption measurement. N_3^* and X_3^* are the total number of ozone molecules (cm^{-2}) and the total ozone equivalent depth (cm STP) respectively in the path.

Total Ozone Amount

Measurements of the total amount of ozone in a vertical atmospheric column, whether made from ground or from satellite-based instruments, depend on optical techniques. Ground-based methods make use of radiance measurements from an external light source like the sun or the moon after the radiation has suffered extinction as a result of atmospheric absorption, molecular scattering, and large particle (aerosol) scattering, all of which are wavelength-dependent. Satellite measurements, on the other hand, are based on the extinction of upwelling radiation whose source is either backscattered solar radiance or infrared emittance from the earth-atmosphere system.

Total ozone observations made from the ground employ either dispersed spectroscopic or optical filter techniques. The standard method presently in use in the International Ozone Network is based almost exclusively on a technique first suggested by Fabry and Buisson (1913, 1921) and then modified by Dobson (Harrison and Dobson, 1925; Dobson and Harrison, 1926). The basic design for the spectrophotometer in current use was discussed by Dobson (1931). Since the development of the original Dobson spectrophotometer there have been many improvements in both the optics and the information evaluation systems of the instrument (see Olafson, 1969; Else, Powell, and Simmons, 1969; Komhyr and Grass, 1972; Raeber, 1973; Westbury, Thomas, and Simmons, 1981). A detailed description of the Dobson instrument and the operating procedure for its use was given by Dobson (1957) and has been made current by Komhyr (1980a).

Optical filters were first used by Fabry and Buisson (1913) for isolating a portion of the ozone-absorbing spectral interval in observations of total ozone. However, the filter technique presently used for most routine measurements is based on a method originally developed by Stair and Hand (1939). This technique has the advantage of being relatively inexpensive (as compared with optical dispersion systems) and is designed to be compact and portable. A basic problem, however, is the need for very frequent calibration as a result of filter degradation and the generally low resolution of the optical filters. A review of the general characteristics of filters currently used for total ozone measurements and the measurement difficulties in their use is given by Gerlach and Parsons (1981).

New ground-based instruments are being developed to supplement the absolute calibration capability of the standard network observations and for use in specific research programs. These instruments are based on high resolution ultraviolet spectrophotometric techniques (e.g., DeLuisi, 1975; Kuznetsov, 1977; Garrison, Doda, and Green, 1979), narrow band filter spectrometers (e.g., Matthews, Basher, and Fraser, 1974; Hanser, Sellers, and Briehl, 1978; Matthews and Fabian, 1981), hetrodyne radiometric methods operating in the infrared (Menzies and Seals, 1977) and millimeter regions (e.g., Shimabukuro, Smith, and Wilson, 1977; Dierich et al., 1981), high resolution infrared spectroscopy (Secroun et al., 1981), UV lidar (Browell, Carter, and Shipley, 1981; Pelon et al., 1981), and others. None of these new techniques is presently involved in a program of routine ozone observations as part of the global system, but they represent important developments in the state of the art of total ozone observations.

Determination of Total Ozone. If the extinction coefficient, including absorption and scattering effects, is independent of the light path, the monochromatic radiance measure at the ground in an ozone absorbing wavelength is

$$L_\lambda = L_{\lambda\infty} 10^{-(a_\lambda X\mu + \beta_\lambda mp/p_0 + \delta_\lambda m')}, \tag{1-2}$$

where

L_λ and $L_{\lambda\infty}$ are the measured radiances at the ground and the top of the atmosphere respectively;

a_λ, β_λ and δ_λ are the decadic ozone absorption (cm^{-1}), molecular scattering for air (atmosphere^{-1}), and aerosol scattering (relative) coefficients;
X is the total amount of ozone in a vertical column (STP);
μ is the relative path length of the solar beam through the ozone layer for
  a spherical atmosphere;
m is the optical path length allowing for refraction through the molecular
  scattering spherical atmosphere (given by Bemporad's formula—see,
  for instance, List, 1968);
m' is the relative slant path through the aerosol atmosphere, assumed to
  be equal to sec ζ, where ζ is the solar zenith angle;
p, p_0 are the station pressure and mean sea level pressure respectively;
μ, m, and m' are significantly different from each other only at high zenith
  angles (i.e., $\zeta \geq \pm 75°$). This difference arises from the different vertical
  distribution of ozone, air molecules, and aerosol particles.

It is convenient to take simultaneous radiance observations at two
nearby ozone-absorbing wavelengths, so that relative rather than absolute radiances can be measured, and to minimize difficulties because of
uncertainties in the aerosol scattering coefficients. The total ozone amount
is then given by

$$X = \left\{ \log \frac{L_{1\infty}}{L_{2\infty}} - \log \frac{L_1}{L_2} - (\beta_1 - \beta_2)mp/p_0 - (\delta_1 - \delta_2) \sec \zeta \right\} \Big/ (a_1 - a_2)\mu, \qquad (1\text{-}3)$$

where subscripts 1, 2 refer to wavelengths of relatively strong and weak
ozone absorption respectively.

Paired wavelengths for ozone observations with the spectrophotometer
were first used by Harrison and Dobson (1925). A set of five standard
wavelength pairs, designated as A, B, C, C', and D, was adopted for use
after July 1957 (the start of the IGY) and in order to further minimize the
troublesome effect of aerosol scattering, the computational procedure
was modified to include observations on double wavelength pairs (Dobson,
1957).

Dobson Spectrophotometer. The Dobson spectrophotometer is a double prism monochrometer that matches radiance measurements at two
different wavelengths about 20 nm apart, in the Huggins bands of the
ozone ultraviolet spectrum (~300-340 nm). An optical wedge is used to
reduce the observed radiance at the weaker absorbed wavelength. A null
setting then allows the relative radiance, and thus the total ozone amount
in the vertical column, to be determined. The wavelengths used for each
measurement pair and the absorption and scattering coefficients adopted

for use with the Dobson spectrophotometer as of January 1968 are given by Komhyr (1980a). These coefficients are modified from laboratory-derived values to be suitable for use with the Dobson instrument. They should not be used for other instrument types without further adjustment to fit the optical characteristics of that particular instrument.

Determination of total ozone by the Dobson method is based on a number of assumptions, the most important of which are as follows:

The major part of the ozone layer is in the lower or middle stratosphere. The relative radiance received at the top of the atmosphere for the different wavelength pairs does not change with time, e.g., during the course of different sunspot cycles.

Absorption in the atmosphere at the observed wavelengths is by ozone alone.

The wavelength dependence of atmospheric aerosol scattering is, for the observed spectral region, linear in first-order differences.

Total ozone observations can be made for solar zenith angles as large as about $80°$. Although the observations are most reliable when viewing the sun directly, useful routine measurements can also be made under other conditions (e.g., viewing the zenith sky under clear or cloudy conditions, or direct observation on the moon). Different wavelength pairs are then combined to fit the different observing condition as appropriate.

Measurement Uncertainties. The Dobson spectrophotometer is the designated standard instrument for total ozone measurements in the global network, and it is used as a calibration base for a number of other total ozone observing systems, including those involving satellite techniques. It is thus important to consider the sources of uncertainty in the Dobson measurements. Many of these sources have been discussed by Dobson in a series of published papers (see Dziewulska-Losiowa and Walshaw, 1975). These involve

1. the possible inaccuracy of the effective ozone absorption coefficients used with the Dobson spectrophotometer and the temperature dependence of these coefficients;
2. the possibility of interference in the radiance measurements due to the presence of trace gases with absorption spectra near wavelengths used in the Dobson instrument;
3. instrumental errors of various types, including the deterioration of the internal optics, the presence of scattered light within the instrument, and the possible drift of the optical wedge characteristics;
4. possible nonlinear variation of aerosol scattering coefficients in the Huggins band;

5. variable extraterrestrial solar radiance ($L_{\lambda\infty}$) at wavelengths used in the Dobson instrument; and
6. the need for correction for polarized light associated with zenith or cloudy sky observations.

As originally pointed out by Ladenburg (1929), the determination of total ozone is very sensitive to the accuracy of the absorption coefficients used in the measurement, and particularly as applied to the observing instrument. As new laboratory data based on improved experimental techniques became available, adjusted ozone absorption coefficients appropriate for the Dobson instrument were applied for use with the spectrophotometer under atmospheric observing conditions. It has been known for some time that total ozone values as determined from different double-paired wavelength observations yielded inconsistent amounts (Dobson, 1963). This inconsistency was partially corrected by the "modified" absorption coefficients adopted for use after 1 January 1968 based on the new laboratory results of Vigroux (1967). The differences for the AD pairs, however, were kept the same. (A review of the different laboratory measurements of the ozone absorption coefficients in the Hartley and Huggins bands since 1953 is given by Klenk [1980].) Although there seems to be only slight inconsistency between total ozone values as determined by different double wavelength pairs used in the Dobson method (Mateer, 1981), it has been suggested by Komhyr (1980b) that the standard AD coefficients may give ozone amounts that are systematically too high by about 5%. Recent preliminary laboratory measurements (Bass and Paur, 1981) suggest that some adjustments need to be made to the absorption coefficients recommended for use with the Dobson spectrophotometer. Indications are that the AD coefficients currently used are too low, but the absolute values of the corrections have not yet been determined.

Absorption coefficient measurements made by Vigroux (1953) were for different temperatures as applicable to atmospheric conditions. However, the recent provisional laboratory results reported by Bass and Paur (1981) also suggest that the temperature dependence of the coefficients is somewhat stronger than that given by Vigroux as used for the A wavelength pair. The coefficients presently used, as modified from the Vigroux measurements, are assumed to be appropriate for a mean stratospheric temperature of -44°C. The calculated total ozone is underestimated for higher average temperatures and vice versa, the error being about 1% for a temperature difference of 10°C (e.g., Walshaw, Dobson, and McGarry, 1971; Powell, 1971). An error of about 2% in the seasonal variation of ozone could arise therefrom since at 25 km the average winter/summer temperature difference at subpolar latitudes is about 20°C. In addition, local perturbations to the stratospheric temperature (e.g., stratospheric warmings) could result in similar error when the absorption coefficients are adjusted to an assumed fixed temperature.

The computed ozone amount may, of course, be in error if atmospheric gases, other than ozone, have absorption bands at wavelengths used for ozone measurements (Dobson, 1963). This "error" could account for some of the resultant inconsistencies discussed above. Although a number of interfering gases have been suggested, only SO_2 seems likely as a possible significant contribution to this discrepancy (Komhyr and Evans, 1980). Estimates by Evans et al. (1981) indicated that with large SO_2 amounts (~5 m atm-cm) the computed ozone amount could be overestimated by about 1-2%. Thus, long term ozone changes could be masked by the effect of long term changes of SO_2, if suitable corrections to the measurements are not made.

In the determination of total ozone from Dobson type measurements it is assumed that the extraterrestrial solar irradiance at the different paired wavelengths is constant over different time periods. Although there may be periodic (e.g., 27 day or solar cycle) or short term aperiodic irradiance variations at individual wavelengths in this spectral region, there are no documented variations that would result in an erroneous change of total ozone by more than a few tenths of a percent if double pair wavelengths are used (WMO, 1981).

Considerable controversy has existed in the past concerning errors in Dobson total ozone measurements that may be associated with aerosol effects (see, for instance, London and Angell, 1982). These errors could arise if the assumption that the scattering coefficients due to aerosols are reasonably linear with wavelength in the spectral region of Dobson observations is incorrect. It has recently been shown, however, (Mateer and Asbridge, 1981) that the aerosol correction is indeed small when the Dobson method is applied to direct sun measurements using the AD double pair wavelengths.

Many improvements have been made in the instruments and operating procedure to minimize the various errors discussed above. Frequent recalibration is made of the optical wedges, since the wedge characteristics can change with time. Advances have been made towards eliminating the internal light scattering problem and correcting for polarizing light from zenith sky and cloudy sky observations. The electronics of the Dobson instrument have been modernized and attempts are presently underway to make the instruments automatic. A modified Dobson instrument was put into operation in 1981 in Boulder, Colorado as part of a program to establish a global automated Dobson station network by spring 1984.

Although direct sun AD observations are recommended whenever possible for Dobson observations, results from zenith blue sky or cloudy sky measurements can be useful provided that appropriate empirical relations between the two observation types are determined. The average long-term precision of the Dobson measurements at a reasonably well-operated station in the Global Network is about 2-3%.

A program has been developed by the World Meteorological Organization for routine intercalibrations of a set of nine regional standard Dobson instruments with World Primary Standard Dobson Spectrophotometer No. 83 operated in the U.S. by the NOAA, Air Resources Laboratory, Boulder, Colorado, designated by the WMO as the World Dobson Spectrophotometer Central Laboratory. The regional standard spectrophotometers are used periodically to calibrate other instruments in the Global Network (Komhyr, Grass, and Leonard, 1981; WMO, 1981).

Brewer Spectrophotometer. A new spectrophotometer was designed by Brewer (1973), based partially on a model constructed by Wardle (1963), as an improvement on the Dobson instrument that would minimize the difficulty of taking observations under different atmospheric conditions. The Brewer instrument, intended as a replacement for or supplement to the Dobson spectrophotometers used in the global ozone observing network, is a grating spectrophotometer designed to measure the solar radiance at five wavelengths in the spectral interval 306-320 nm with a resolution of 0.6 nm (the Dobson spectral resolution is about 1.0-1.5 nm). Calculations of total ozone with the Brewer instrument follow the Dobson method using paired wavelengths as given in Eq. (3). The radiance at each wavelength, however, is measured by a pulse counting system that eliminates the need for the troublesome optical wedge. The instrument has internal calibration capability and is still small enough to be portable. Through use of a polarizing prism, reasonably accurate observations on a clear or cloudy zenith sky can be made without the need of an empirical chart (Brewer and Kerr, 1973). In addition, observations in the wavelengths adopted for use with the Brewer instrument can be combined to provide information for determining the column abundance of SO_2. This information can be used as a correction factor in total ozone observations taken in heavily polluted air. Results of extensive field programs have shown an rms difference in computed total ozone amounts of less than 1% when Brewer and Dobson AD measurements are compared (Kerr, McElroy, and Olafson, 1981; Parsons, Gerlach, and Williams, 1981).

M-83 Filter Ozonometer. The principal optical filter instrument in use in the global ozone network since the start of the IGY is the M-83 Filter Ozonometer (Khrgian, 1975). (A filter instrument designed by

Vassy and Rasool [1959] has also been used for some routine observations.) The M-83, employed chiefly in the USSR and Eastern Europe, was originally based on the measured ratio of the solar irradiance received in three relatively broad spectral intervals in the Huggins bands. The instrument was later modified to employ two filters for total ozone measurements centered at 304 nm and 330 nm (Guschchin 1974, 1977). Total ozone is then derived from the measurements of the ratio of the received radiance at the two different wavelength bands (see Eq. 3), at different instrument temperatures and solar zenith angles, with the aid of a calibration nomogram that is provided for each station using an M-83. The major error arising from the use of the M-83 involves the standardization and calibration of the broad bandpass optical filters and the large field of view of the instrument ($6°$) (Osherovich, Rozinskiy, and Yurganov, 1969; Bol'shakova, 1976). Relatively large errors (up to 20%) in the reported total ozone amount occur at low solar elevations, primarily because of the shift of the effective central wavelength of maximum filter transmission as a function of the solar zenith angle. Significant errors also occur when observations are made in the presence of large aerosol concentrations (see, for instance, Bojkov, 1969; Bol'shakova, 1979). Routine direct sun or zenith blue sky observations made with the M-83 are much more reliable than those taken through cloudy skies (Szwarc, 1981).

As a result of experimental and theoretical studies, (see, for instance, Osherovich, Rozinskiy, and Yurganov, 1969), improved filters and data processing techniques were introduced for use with the M-83, and measurements made with the M-83 after 1972 are more consistent with those made with the Dobson spectrophotometer (WMO, 1981). However, recent intercomparisons between M-83 and Dobson spectrophotometers confirm that there is still some air mass dependence of the difference between total ozone measured with the two instruments. During periods of relatively high sun (summer), M-83 derived total ozone values are slightly less than those observed with the Dobson instrument; during periods of low sun (winter), the reverse is true (Szwarc, 1981). Computation of the rms difference between an improved M-83 and the Dobson spectrophotometer for a long series of simultaneous measurements was 3.8% (Parsons, Gerlach, and Williams, 1981). The accuracy of the presently used M-83 in the observing network is estimated to be about ±3-5% for direct sun or zenith blue sky observations.

Satellite Techniques. The two principal techniques used for satellite measurements of total ozone involve observations of solar ultraviolet radiation which has been backscattered from the earth and atmosphere (Dave and Mateer, 1967; Mateer, Heath, and Krueger, 1971), or observations of infrared emission at 9.6 μm from the ground and atmosphere (Sekihara and Walshaw, 1969; Prabhakara, 1969; Prabhakara et al., 1970). The basic methodologies used in these techniques are summarized by WMO (1981).

The backscatter ultraviolet (BUV) method measures the upwelling solar radiation that has been absorbed by ozone along the path of the solar beam and then scattered back to the satellite by the earth and atmosphere. The principles of the method are very similar to those involved in the Dobson technique. Details of the observational technique and method for deriving total ozone data from the observations are described by Heath, Mateer, and Krueger, 1973; Klenk et al., 1982. In the case of BUV observations, some difficulties have been encountered largely because of instrument degradation (Fleig et al., 1981) and uncertainty of the correct "effective" ozone absorption coefficients for the wavelengths used in the BUV instrument (Klenk, 1980), and because the presence of cloud layers requires some approximations for contributions of tropospheric ozone to the total amount (Klenk et al., 1982). As a result, improvements have been made on the original design of the BUV instrument (Heath et al., 1975) and procedures for determining total ozone amounts with the new instruments (Fleig et al., 1982). These have been incorporated in the Solar Backscatter Ultraviolet (SBUV) and the Total Ozone Mapping Spectrometer (TOMS) instruments both flown on the Nimbus-7 satellite.

The total amount of ozone in a vertical column above the earth's surface or cloud top can also be determined from satellite observations of the upward-directed infrared radiance in the 9.6 μm ozone band provided that the vertical temperature distribution is known. The method for data reduction requires the solution of an integral equation whose kernel function is not very well determined (see, for instance, Prabhakara et al., 1976). As a result, a set of approximations are used whereby total ozone, as observed from ground-based Dobson measurements, is correlated with clear sky radiances observed by the satellite instrument. This method has been used with the Nimbus-3 and Nimbus-4 IRIS system (Prabhakara, Rodgers, and Solomonson, 1973), the Defense Meteorological Satellite Program—Multichannel Filter Radiometer (MFR) (see Lovill et al., 1978; Luther and Weichel, 1981), the TIROS Operational Vertical Sounder (TOVS) (Smith et al., 1979), METEOR 28 (see Feister and Spänkuch, 1981; Feister, 1980). The infrared technique is sensitive to the independently determined mean temperature of the lower and middle stratosphere and the initial assumed vertical ozone distribution. Uncertainties in total ozone determination due to this technique can be as large as $\pm 10\%$ (WMO, 1981).

Comparison between satellite measurements (BUV) and quasi-synchronous co-located Dobson or M-83 total ozone observations has

Table 1-1. Satellite Total Ozone Measurements

Satellite	Instrument	Observation Period
Nimbus-3	IRIS	4 April–July 1969
Nimbus-4	IRIS	April 1970–January 1971
Nimbus-4	BUV	April 1970–July 1977
AE-5	BUV	November 1975–May 1977
DMSP	MFR	March 1977–June 1979
Nimbus-7	SBUV	November 1978–
Nimbus-7	TOMS	November 1978–
Tiros N	TOVS	November 1978–February 1981

been discussed by Miller et al. (1978) for the period April 1970 to April 1971 and by Fleig et al. (1981) (see, also, WMO, 1982) for the 7-year period April 1970 to May 1977. The average annual difference (Dobson-BUV) increased slightly during the 7-year interval; this shows the effect of the degradation of the diffuser plate on the BUV instrument. The correlation between individual satellite and Dobson values was slightly higher than +0.9 and fairly constant during the entire period. However, the correlation between the satellite and M-83 measurements was less than +0.7 before 1974 but dramatically increased to almost +0.9 after 1974. It is clear that the improvements made in the M-83 instrument added significantly to the quality of the ground-based global network.

Comparisons have also been made for total ozone data as derived from infrared (IRIS, TOVS) and ultraviolet backscattered (BUV, SBUV) satellite measurements, and values observed with Dobson instruments (Prior and Oza, 1978; Crosby et al., 1981). The results indicated that total ozone retrieval using infrared methods does not adequately reproduce the BUV or ground-based observed ozone patterns at subpolar latitudes.

A list of the different satellites providing total ozone data and the observational period for which they are available is given in Table 1-1 (WMO, 1981).

Vertical Distribution

The observed abrupt ultraviolet termination of the solar spectrum was attributed by Hartley (1880b) to absorption by atmospheric ozone. As a result, direct measurements of the extinction of ultraviolet radiation as the sun descended toward the horizon led Hartley (1881b) to conclude that the ozone concentration was in larger proportion in the upper atmosphere than near the earth's surface. Early observations resulting from a suggestion by Fabry and Buisson (1913) placed the ozone maximum at about 50 km (Strutt, 1918), but it was shown by Götz and colleagues using a new technique discussed below (Götz, 1931; Götz, Meethan, and Dobson, 1934) that the average level of the ozone maximum is closer to about 22 km. Subsequent observations confirmed that this maximum ozone level varies with season, latitude and meteorological conditions (e.g., Dütsch, 1978; London and Angell, 1982).

Many observing techniques are presently being used for routine measurements of the vertical ozone distribution: see, for instance, WMO (1981). The observing methods are either indirect (remote) or direct (in situ). All remote methods employ optical techniques operating at various wavelengths. The observations are generally ground based or are made from satellite platforms. The in situ methods are optical, electrochemical, or chemiluminescent and are made from balloon-borne or rocket-borne instruments and, for some extended programs, from aircraft platforms. A third technique gives the local ozone concentration derived from optical measurements of total ozone above the platform as made from vertically rising balloons or rockets.

Remote Systems

Ground-Based Methods. The standard ground-based method for determining the vertical ozone distribution is the Umkehr (Götz) technique, which makes use of spectrophotometric observations to determine the mean ozone concentrations in nine layers from the ground to about 48 km. The technique is based on a series of measurements of the radiance ratio at two narrow spectral intervals, one strongly and one weakly absorbed by ozone, of light scattered down from the blue zenith sky as the sun approaches the horizon. The name Umkehr is derived from the form of the plotted curve of the log radiance ratio as a function of the solar zenith angle, which decreases to a minimum at a zenith angle of about 85° and then increases. If observations are made as the sun sinks below the horizon, a second reversal occurs in the curve when the sun is at zenith angle of about 95° (see, for instance, Ramanathan, Angreji, and Shah, 1969). The Umkehr effect results from the dependence of the contribution functions for the downward scattered radiance on the optical depth (due to scattering and absorption) in the atmosphere along the path of the radiation beam. As the solar zenith angle increases, the center of the contribution function moves upward from near the troposphere to well above the height of the layer of maximum ozone density (approximately 22-25 km). This upward shift is faster the stronger the absorption, and is faster when the center of the contribution function for each wavelength is below rather than above the height of ozone maximum.

If it is assumed that the radiation suffers extinction due to absorption and Rayleigh scattering as the radiation follows the slant solar beam and is then scattered vertically downward, the ratio of the radiances received at the ground at two different wavelengths is, from Eq. (2) (see, for instance, Craig, 1965)

$$\frac{L_1}{L_2} = \frac{L_{1\infty}}{L_{2\infty}} \frac{K_1}{K_2} \frac{\int_0^\infty \rho \tau_1 (0 - z) \, \tau_1 (z - \infty) \, dz}{\int_0^\infty \rho \tau_2 (0 - z) \, \tau_2 (z - \infty) \, dz},$$ (1-4)

where the transmittances are defined by

$$\tau_i(0 - z) \equiv 10^{-\int_0^z |a_i X(z) + K_i \rho| \, dz}$$

$$\tau_i(z - \infty) \equiv 10^{-\int_z^\infty |a_i \mu' X(z) + K_i \rho \mu'| \, dz},$$

and where K_i is the mass scattering coefficient over each wavelength interval ($K_i \rho \simeq \beta_i \, mp/p_0$), $X(z)$ is the ozone density variation with height (STP), and μ' is the secant of the angle between the solar beam and the local vertical at height z.

The vertical ozone distribution $X(z)$ is obtained as a solution to the inverse problem (Eq. (4)), where an assumed vertical distribution and its statistical properties are used as input information and the calculated distribution is derived from successive approximations.

In practice, the current standard Umkehr method involves observations of the C wavelength pair taken either in the morning (from just before sunrise to a solar zenith angle of about 60°) or in the late afternoon (from a solar zenith angle of 60° to just after sunset). The observing procedures and method of evaluation of the ozone distribution from Umkehr observations are discussed by Mateer and Dütsch (1964).

Ozone information derived from Umkehr observations is more accurate for the upper layers, above the ozone maximum, than below. The measurements suffer from the same sources of error as described above in the case of total ozone measurements made with the Dobson spectrophotometer. A major source of uncertainty in the ozone profiles derived from Umkehr observations is due to the inversion algorithm used in the method. Also, since only a single wavelength pair is used for the computations, the assumption of single scattering in a pure Raleigh atmosphere is somewhat tenuous, particularly in the presence of significant stratospheric aerosols. It has been shown that large stratospheric aerosol concentrations, as in the case of volcanic eruptions, cause the stratospheric ozone concentration to be underestimated (DeLuisi, 1979; WMO, 1982).

The ozone distributions as derived from Umkehr measurements generally contain only four independent pieces of information (Mateer, 1965). This is a result of the broad contribution function of the downward-directed radiance and the high correlation of ozone concentrations in different adjacent layers. Because the Umkehr technique is thus limited in its information content, details of the vertical distribution are severely suppressed (e.g., Craig, 1976). The Umkehr observations tend to underestimate systematically the ozone concentration at the level of the ozone maximum and to overestimate the concentration just above. All Umkehr profiles are uniformly computed at the World Data Center in Canada based on a standard evaluation technique developed by Mateer and Dütsch (1964).

A regular Umkehr observation takes about 3 hours for a full set of measurements, during which time upper-level ozone and zenith sky conditions may very well change. A shortened method has therefore been suggested to reduce the time necessary to less than 1 hour for an observation (DeLuisi, 1978, 1979). This method is based on clear zenith sky observations taken on different wavelength pairs at a limited number of solar zenith angles in the early morning or late afternoon. A system of nonlinear equations involving observations at three Dobson wavelength pairs (A, C, and D) and six zenith angles (from 80° to 89°) is then solved by an improved inversion technique (Mateer and DeLuisi, 1981). Comparisons show only a small rms difference above 10 mb when the ozone profiles are derived from standard and short methods. Results for the upper levels appear to be less sensitive to aerosol scattering effects. Random errors for both methods are estimated to be approximately ±5-8% in the stratospheric layers above 15 mb (WMO, 1982).

Methods for determining the vertical ozone distribution from ground-based observations have also been developed using infrared (Goody and Roach, 1956; Vigroux, 1959), microwave (Shimabukuro, Smith, and Wilson, 1975; Kunzi and Fulde, 1977; Parrish, deZafra, and Solomon, 1981), and Lidar (Pelon et al., 1981) techniques. The infrared method is based on observations of atmospheric emission of 9.6 μm and has been discussed by Walshaw (1960), Brewer et al. (1960), Dave, Sheppard, and Walshaw (1963); Vigroux (1971), and others. It has been found that some useful information of the ozone concentration could be obtained by this method but it is limited generally to the upper troposphere and lower stratosphere (see also Secroun et al., 1981).

Microwave measurements in the rotational spectral region (100-130 GHz), on the other hand, can provide ozone profiles in the upper stratosphere and mesosphere. These measurements can be made during the day or night and, moreover, under conditions of overcast skies. The microwave techniques are not presently suitable for routine observations but hold great promise for future ozone measuring systems.

Satellite Platforms. The use of a satellite (reflected light from the moon that has first been refracted in the earth's upper atmosphere at the time of a lunar eclipse) to determine the vertical ozone distribution was proposed in 1931 by Götz (1931; see also Penndorf, 1948) and experimentally carried out by Barbier, Chalonge, and Vigroux (1942) and by Paetzold (1950, 1955). A modification of this method by Venkateswaran, Moore, and Krueger (1961) was applied to visible sunlight reflected from an artificial satellite (Echo I) as it was occluded by the earth's shadow. Analogous observations of solar ultraviolet radiation made from a polar orbiting satellite during sunrise and sunset were reported by Rawcliffe et al. (1963).

Remote ozone observing systems from satellites have been of three general types. The first, originally proposed by Singer (1956), is in principle analogous to the Umkehr method in that it uses a nadir or limb viewing instrument that measures the reflected backscattered ultraviolet radiance from different layers in the stratosphere and mesosphere (see Iozenas et al., 1969; Anderson et al., 1969; Heath, Mateer, and Krueger, 1973; Thomas et al., 1980). The second type observes infrared emission, usually at 9.6 μm, from different layers of the atmosphere as the instrument goes through a limb scanning routine (e.g., Gille et al., 1975). A third system measures the absorption in the ultraviolet or visible ozone bands as a star (generally the sun) is occulted (e.g., Hays and Roble, 1973; McCormick et al., 1979).

Backscatter Measurements. One of the first set of observations of backscattered ultraviolet radiation used to derive the vertical ozone distribution was made from the nadir viewing instrument on the OGO-4 satellite. The measured upwelling radiance in the spectral interval 255-310 nm was used to solve a set of inverse equations analogous to Eq. (4) (Anderson et al., 1969). The method of solution was based on the techniques suggested by Twomey (1965). Optical rocketsonde observations were used to calibrate the ozone measurements, and ozone distributions were computed for the middle and upper stratosphere (35-55 km) with an estimated accuracy of about 10%.

The BUV and SBUV observations are based on the inversion of upwelling atmospheric backscattered radiance data at 12 wavelengths in the Hartley-Huggins bands (255.5-340 nm). The inversion algorithm used for ozone retrieval has been discussed by Mateer, Heath, and Krueger (1971) and Klenk et al. (1982). The main sources of error in the BUV observations are due to uncertainties in the ozone absorption cross sections (as discussed earlier), algorithm sensitivity, and instrument calibration. There is, in addition, contamination of the radiance data at the longer wavelengths in the presence of large stratospheric aerosol concentrations.

In both the BUV and SBUV instruments a diffuser plate is used to reflect sunlight through the optical system. However, as mentioned earlier, the diffuser plate on the Nimbus-4 BUV instrument suffered degradation soon after launch, which resulted in additional uncertainty in the BUV profiles. Modifications of the system design were made on the SBUV instrument to reduce this problem (Heath et al., 1975). It is estimated that the BUV-type instruments have an overall 5-10% accuracy for derived ozone profiles in the middle and upper stratosphere (1-20 mb) (WMO, 1982). As in the case of the Umkehr measurements, the height resolution of nadir UV observations is limited to about 6-8 km.

The Solar Mesosphere Explorer (SME), a polar orbiting satellite designed to measure the ozone profile in a limb-scanning mode, was launched in October 1981. SME observations are made using a double channel programmable Ebert-Fastie spectrometer to measure limb backscattered ultraviolet sunlight at two wavelengths in the Hartley band. The ozone profile can be derived using the UVS instrument through the height interval 40-70 km and, because of the narrow field of view, with a height resolution of about 3.5 km (Thomas et al., 1980; Rusch et al., 1983).

Limb Observations.   Ozone profiles can also be derived from satellite measurements made from limb-scanning instruments observing emitted atmospheric infrared radiation or absorbed solar visible radiation.

One type of observation measures the radiation emitted at 9.6 μm by ozone in a column along a tangent path in the atmosphere. Since the emittance depends on the local temperature in addition to the ozone concentration, observations are also made at two wavelengths in the 15 μm CO_2 band to derive the temperature distribution. Limb measurements made with instruments on Nimbus-6 (Limb Radiance Inversion Radiometer) and Nimbus-7 (Limb Infrared Monitor of the Stratosphere) have been used for calculation of ozone profiles for the height interval from about 20-65 km with a resolution of about 3 km. However, the instruments require continued internal cooling to maintain a reasonably high signal-to-noise ratio and so far have only limited lifetimes (approximately 6 months) in otherwise successful operations (see Gille and House, 1971; Gille et al., 1981). The SME instrument package also contains a near infrared (NIR) airglow spectrometer that measures the daytime limb radiance at 1.27 μm that results from emission by excited molecular oxygen (O_2 $^1\Delta_g$), a product of photodissociation of ozone (Thomas et al., 1983a). The 1.27 μm radiance measurements provide global routine observations of the ozone distribution in the region 50 to 90 km.

Long path ultraviolet absorption measurements have been made using stellar occultation techniques. Satellite observations of a target star in the wavelength interval 255-310 nm were used by Riegler et al. (1976) and Riegler et al. (1977) to compute the ozone density at heights of 47-114

Table 1-2. Satellite Vertical Ozone Profile Measurements

Satellite	Instrument	Observation Period
OGO-4	BUV type	September 1967–January 1969
Nimbus-4	BUV	April 1970–May 1977
Nimbus-6	LRIR	June 1975–January 1976
OSO-8	UVMCS	June 1975–September 1978
AE-5	BUV	November 1975–May 1977
Nimbus-7	LIMS	October 1978–May 1979
Nimbus-7	SBUV	November 1978–
Atm. Exp. Miss.	SAGE	February 1979–
SME	UVS	October 1981–
SME	NIR	October 1981–

km. Analogous observations on OSO-8 employed a multichannel UV photometer to observe the sunrise and sunset extinction of solar radiance near the MgII doublet (280.3 nm and 279.6 nm) at the long wavelength portion of the Hartley bands. These data provided information on the ozone distribution in the mesosphere from 50-75 km (Millier et al., 1979; Millier, Emery, and Roble, 1981). Observations made from the SAGE satellite of the limb extinction of solar radiation in the Huggins bands (at 0.65 μm) have been used to derive the ozone distribution from the upper troposphere to about 50 km (McCormick et al., 1979). The estimated uncertainty of the SAGE data is about ±2-5%.

It has been suggested (see, for instance, Pokrovsky and Kaygorodtsev, 1978; Timofeyev and Biryulina, 1981) that the use of multi-wavelengths in the infrared and a combination of infrared and UV spectral observations can considerably enhance the information content of both limb and nadir viewing measurements in deriving the vertical ozone profile.

A list of satellites providing routine profile data is given in Table 1-2 (after WMO, 1982).

In Situ Measurements. In situ measurements of ozone in the free air have been made using balloon, aircraft and rocket platforms, each of which is capable of providing either optical or chemical type observations. Direct optical measurements were first made from balloon-borne instruments (Regener and Regener, 1934) and later from rockets (see, for instance, Durand, 1949; Johnson, Purcell, and Tousey, 1951). Wet chemical techniques, used first on aircraft (Kay, 1954; Kay, Brewer, and Dobson, 1956), were modified for automatic balloon measurements (Brewer and Milford, 1960). Balloon observations using a dry chemical method (Regener, 1960, 1964) were adapted for use in providing high level ozone data from rockets (Hilsenrath, Seiden, and Goodman 1969).

Electrochemical Methods. The current standard method of taking direct routine upper air ozone observations is through use of balloon-borne electrochemical ozonesondes of either the Brewer/Mast type (e.g., Attmannspacher and Dütsch, 1970) or the ECC type (Komhyr, 1969). The Brewer/Mast ozonesonde is based on a technique for the continuous measurement of the ozone concentration developed by Paneth and Gluckhauf (1941). Air containing ozone is pumped through a potassium iodide solution in which a platinum gauze cathode and a silver wire anode are immersed. The ozone in the air reacts with the potassium iodide in the solution to free iodine. When a small potential is applied between the electrodes a current is produced, if free iodine is present, proportional to the amount of ozone reacting with the solution. In the measuring process, iodine reacts chemically with the silver anode and is thus effectively removed from the solution.

A slightly different type of electrochemical ozonesonde (Komhyr, 1969; Komhyr and Harris, 1971) consists of two platinum mesh electrodes submerged in buffered potassium iodide solutions in separate cathode and anode chambers. The chambers are electrically linked by an ion bridge that serves to keep the two electrolytic solutions separated. The electromotive force of the system is derived from the difference in anode and cathode electrolyte concentrations (about 80:1). Thus, the sensor functions without an externally imposed polarizing voltage. As in the Brewer/Mast instrument, an electric current is developed proportional to the rate at which ozone enters the sensor and reacts with the solution. Both instruments have relatively fast response times and therefore can measure details of the ozone distribution as the balloon-borne instrument ascends. Because of the solution volatility and difficulties with the pump operation, particularly at low pressures, the electrochemical instruments currently have an effective height limit of about 30 km. However, a program involving an improved carrier balloon and a more reliable pump operating at high altitudes was developed in 1981 to extend the ceiling of the ECC sondes to 40 km. Although the instruments are, in principle, absolute, it has been found that a correction needs to be applied to the measurements to make the measured vertically integrated amounts compatible with the total amount as observed with the Dobson spectrometer (Attmannspacher and Dütsch, 1970, 1981). Some modifications have been made in the Brewer/Mast sonde for use in India (Sreedharan, 1968) and in the German Democratic Republic (Sonntag, 1976). The ECC sonde was modified somewhat for use in Japan (Kobayashi and Toyama, 1966).

Ozone profile measurements made with different chemical techniques have been compared by Hering and Dütsch (1965) and in a WMO-sponsored international intercomparison program by Attmannspacher and Dütsch (1970; 1981). In the most recent intercomparison, involving five improved wet chemical ozonesonde types, it was concluded that the results obtained from the different sondes could be usefully combined in a global ozone profile network. Intercomparisons between electrochemical sondes and Umkehr measurements (e.g., Dütsch and Ling, 1969; Kulkarni and Pittock, 1970; WMO, 1982) showed that in general the Umkehr derived values are higher than those determined from ozonesonde measurements at levels of 15-60 mb. As mentioned earlier, this results largely from the averaging technique used in processing the Umkehr observations. The correlation between Umkehr and ozonesonde data is lowest at levels of 15-30 mb. The estimated accuracy of the electrochemical sondes is better than 10% (WMO, 1981).

Chemiluminescent Methods. When air containing ozone is brought into contact with a luminol or a rhodamine-B dye, luminescence will result (Bernanose and René, 1959), and photons are then emitted in proportion to the ozone concentration in the air stream. The effect of this reaction was used by Regener (1964) to design a relatively simple chemiluminescent device to measure the ozone concentration in the free air. The instruments, however, although calibrated before each flight, were subject to large errors because of unstable disk characteristics and sensitivity to air flow variations. The Regener ozonesondes were flown fairly extensively in a North American network during the period 1963-1965 (Hering and Borden, 1967), but because of difficulties encountered in quality control they are no longer being used. The chemiluminescent system is not height-limited and has been successfully adapted for rocket observations (Hilsenrath, Seiden, and Goodman, 1969; Konkov, Kononkov, and Perov, 1981). A new balloon-borne technique, whereby the instrument measures ozone by means of a chemiluminescent reaction of ozone with vaporized ethylene, was recently used effectively for observations at heights of about 42 km (Aimedieu and Barat, 1978; Aimedieu et al., 1981).

Optical Methods. Optical methods generally make use of spectroscopic observations of solar radiation in the Hartley-Huggins bands. The total amount of ozone above the instrument can be determined from the measurements as the balloon or rocket ascends through the atmosphere. The vertical ozone distribution is then obtained by differentiation of the calculated total amount as a function of height.

The first direct optical measurement of the ozone distribution in the stratosphere was that of E. Regener and V. H. Regener (1934) who recorded the solar ultraviolet signal on a spectrograph on board an unmanned balloon as it ascended to 29.3 km. The observation verified earlier indirect (Umkehr) measurements of Götz (1931) that the ozone density maximum was at a height of about 25 km.

The most extensive program of optical ozonesonde observations was that conducted by Paetzold and associates during a period of about 20 years (1951-1972) (Paetzold, 1973). These observations, often up to a height of 30 km, were not, however, homogeneous in space or time, and there are no current plans for the use of this technique in a routine ozone monitoring program.

Profiles of stratospheric ozone have also been obtained recently from balloon measurements of atmospheric limb thermal emission at millimeter wavelengths (Waters et al., 1981) and through the use of ultraviolet LIDAR observations (Heaps et al., 1981). In the latter case, the ozone concentration is derived along the optical path of the returned LIDAR signal from the measured differential absorption at two wavelengths, one strongly and one weakly absorbed, in the Hartley-Huggins bands (see also Browell, Carter, and Shipley, 1981; Pelon et al., 1981).

A self-contained optical system using a flash-tube as a light source in tandem with a spectrograph on a carrier balloon was developed by Regener (1954) to measure the stratospheric ozone profile. This principle was used to design a continuous ozone measuring technique that employs a mercury vapor lamp as an emitting light source centered at 253.7 nm (the Dasibi ozone photometer). For a known ozone absorption cross section, the attenuation of the light beam, as measured by the Dasibi photometer, is a function only of the ozone concentration in the fixed path. Details of the Dasibi instrument are given by Bowman and Horak (1974). Although originally designed to measure the ozone concentration near the ground, the Dasibi has been modified for use with balloons (Robbins and Carnes, 1978) and on aircraft in connection with the Global Atmospheric Sampling Program (GASP) (see, for instance, Falconer and Holdeman, 1976). It is in principle an absolute instrument with height and time resolutions approximately equivalent to that obtained by the wet chemical techniques. Estimates of the instrument errors give an upper limit of $\pm 4\%$ except for errors possibly due to ozone loss at the tube walls, which might vary between 0-30% at 40 km (Ainsworth, Hagemeyer, and Reed, 1981).

Results of recent intercomparison programs in the U.S. and in France, involving the various in situ measurement techniques described above, have been discussed in WMO (1982). In the lower and middle stratosphere the rms difference of the ozone density, as derived from the different instrumental techniques, is about 15-20%. Similar intercomparisons are planned as part of the continued international effort to improve the performance of both routine and research ozone observing systems.

Rocket Observations. Different observing techniques have been employed for ozone profile measurements from rocket platforms. One, based in principle on the method used with the first optical rocket flown in 1946 (Johnson, Purcell, and Tousey, 1951), uses the sun or moon as a light source and measures the absorption of incoming solar radiation at one or more ultraviolet wavelengths as the rocket ascends. This method can be employed for daytime or nighttime measurements (e.g., Carver, Horton, and Burger, 1966; Tohmatsu, Ogawa, and Watanabe, 1974; Subbaraya and Lal, 1978; Lean, 1982). Similar rocket observations have been made in the USSR (see, for instance, Kuznetsov, Chizhov, and Shtyrkov, 1975; Brezgin et al., 1977). Another method is based on a technique first designed by Kulcke and Paetzold (1957) for use on balloons and later developed by Krueger (1965) and Krueger and McBride (1968) for use on rockets. This method measures the downward-directed solar radiance received on a horizontal diffuser plate, by means of filter photometers, as the payload descends after ejection from the rocket. Interference filters such as those used in the optical rocketsondes tend to have relatively short shelf lives and need careful quality control of their filter characteristics before each rocket flight.

As mentioned earlier, rocket-borne chemiluminescent instruments are also being used to measure the vertical ozone distribution in the stratosphere and mesosphere (20-65 km). The system is the same as was used for the Regener type ozonesonde in that the measurement principle employs the chemiluminescent reaction of ozone with rhodamine-B dye. The new rocketsonde has improved stability of the chemiluminescent detector and better calibration procedures before each flight than were employed previously (Hilsenrath and Kirschner, 1980). It is estimated that the absolute accuracy of the measurements is $\pm 12\%$. The advantage of this technique is that the measurements can be made during the night and, at high latitudes, during winter.

Excited molecular oxygen resulting from photodissociation of ozone in the mesosphere and lower thermosphere emits infrared radiation at 1.27 μm. This radiation can be detected by a vertical or limb viewing airglow photometer mounted on a rocket (or satellite). Such observations have been discussed by Reed (1968) and by Evans and Llewellyn (1972) and are currently being made from SME satellite limb observations (Thomas et al., 1983a).

Although there is no current routine global rocketsonde network, the results of the U.S. rocketsonde observations for the period 1965-1971 were discussed by Krueger (1973). The Japanese rocket flights were summarized by Tohmatsu (1977) and by Ogawa and Watanabe (1981) (see also Ilyas, 1981). An international ozone rocketsonde intercomparison was held at the Wallops Flight Center Facility in the fall of 1979 with five countries participating and involving six different optical and chemiluminescent rocketsonde types (see Sundararaman et al., 1981). Only a preliminary evaluation of the results of the intercomparison is currently available (WMO, 1982; Lean, 1982).

The characteristics of ground-based, balloon-mounted and satellite mounted instruments used in the Global Observing Network for measurements of total ozone and its vertical distribution are summarized in Appendix C of the report by WMO (1982). All observed data from the global network are published bimonthly by the World Ozone Data Center, Downsview, Ontario, Canada. Satellite data are archived and are available at the National Space Science Data Center in Greenbelt, Maryland.

THE OBSERVED DISTRIBUTION

Total Amount

There were very few routine measurements of the total ozone amount during the first quarter of the twentieth century. However, as a result of the prodigious efforts of G. M. B. Dobson and his colleagues, a coordinated program was established that provided sufficient observations during the period 1925-1928 to permit a description of the basic pattern of seasonal and latitudinal total ozone variations as well as the day-to-day changes associated with meteorological activity in middle latitudes (Dobson, Harrison, and Lawrence, 1929; Dobson, 1930). The present ground-based international ozone network was established largely as a result of the extraordinary cooperative efforts of the IGY Program, which started in July 1957. During the past 25 years, ground based and satellite observations have provided details that confirm many aspects of the earlier picture of the ozone distribution and its variations so well described by Dobson and his co-workers over 50 years ago. The recent observations have also provided the data base currently used in an attempt to unravel the complex problem of long-term variations. The major contribution of the observational program over the past 15 years, however, has been the emergence of a nearly complete three-dimensional description of the ozone distribution (discussed in a later section). An excellent summary of published studies of total ozone and ozone profile satellite data has been published in Appendix C of WMO (1982).

Ground-based Observations.   Since the start of the IGY, total ozone measurements have been made at over 170 different locations, but only about half of these stations have been in operation on a regular routine basis for more than a few years. At one time or another about 130 Dobson-type spectrophotometers were built, of which about 65 are currently being used. (Many were shifted from one place to another after only relatively short periods, of the order of a few months, of observations.) About 25 M-83 filter instruments are presently in use, mostly in the USSR and Eastern Europe.

Ground-based observing stations operate only over land areas. (There are occasional shipboard observations, but these are not part of the routine network.) Most of the observations are in midlatitudes, and slightly more than 75% are in the Northern Hemisphere. As a result of the geographic distribution of observing stations, there is potential for a strong bias in the averaged total ozone amounts (zonal, hemispheric, and global) as calculated from the observed station values. The geographic bias results from the following: since total ozone is inversely correlated with the pressure pattern in the upper troposphere and lower stratosphere (e.g., Craig, 1965; Dütsch, 1969), year-to-year shifts in the long wave pattern of the general circulation could produce a fictional zonal or hemispheric averaged ozone variation as a result of the fixed location of the observing stations. Area-weighted hemispheric averages are particularly sensitive to the data derived from observations at low latitudes, where there are very few stations, but where the time and longitude variations are quite small. In addition, measurements at polar latitudes are not made during the winter. (This restriction holds true for most total ozone satellite observations as well.) There are no quantitative estimates available for the bias in the averaged data introduced by these factors. But it has been suggested, from indirect evidence, that for hemispheric averages it is probably not more than a few percent (e.g., Moxim and Mahlman, 1978; Miller et al., 1978; London and Ling, 1981).

In addition to the geographic bias, there is some uncertainty in the averaged ozone measurements related to a "meteorological bias" associated with the presence of dense clouds at the time of the observations (see, for instance, Dobson, 1957; Brewer and Kerr, 1973). Clouds have two effects on the ozone measurements: the first, which results from multiple scattering in the cloud, produces an apparent increase in ozone; the second, associated with extensive convective clouds, may involve a real increase of ozone. The first effect can be corrected for by a suitable adjustment of the instrument optics. The second, however, introduces a real bias to the averaged ozone values. In addition, at some stations, ozone observations are not taken when there is heavy overcast or rain present. Since these conditions generally occur during cyclonic conditions when total ozone is higher than average, observation limited to clear or partly cloudy days would bias the data sets towards lower ozone values. However, it has been estimated that the "cloud bias" is much less than the sampling error that would result from a sharp decrease in the effective density of observations (Moxim and Mahlman, 1980). Thus, long term variations could be the result of climate variations of observing conditions rather than real variations of ozone amounts (e.g., Greenstone, 1978). Despite the uncertainties and data selectivity just discussed, the information content of the averaged total ozone observations from the global ground based network is still quite high, as can be seen when comparisons are made with satellite-derived data.

The average total ozone amount is close to 300 m-atm cm when averaged over the globe. (This amount is equivalent to 8.1×10^{18} molecules cm^{-2} in a vertical column above sea level. The total amount varies geographically and seasonally. In addition, there are local and interannual changes associated with transient meteorological systems and large scale fluctuations in the year-to-year patterns of the general circulation, particularly in the lower stratosphere. Total ozone changes may also be affected by long term solar variability, variations in the concentration of related atmospheric trace constituents, or other geophysical variables.

The ozone partial pressure is a maximum in the middle stratosphere. On the average, about 60% of the total ozone is found in the layer 15-30 km, and variations in total ozone largely reflect variations in the ozone concentration at these heights. Since the photochemical relaxation time for ozone below 30 km is longer than the relaxation time due to stratospheric transport processes (see, for instance, Crutzen and Ehhalt, 1977), total ozone variations on time scales of the order of months are strongly influenced by changes of the lower and middle stratospheric circulation patterns. At some stations, particularly in midlatitudes, there are significant changes within a single day, from day-to-day within a single month and, notably during the spring, from year to year (see, for instance, Dütsch, 1974). During the spring the monthly total ozone amount occasionally increases to above 550 m-atm cm over northern Canada or over central Siberia and could reach as high as 600 m-atm cm on a single day. These extreme values are not observed in the Southern Hemisphere where satellite or ground based measurements seldom report monthly averages of more than 450 m-atm cm of ozone.

The long term annual average distribution of total ozone for the 23-year period (1958-1980), as determined from the ground based global network, is shown in Figure 1-1 (*A*). The filled circles and triangles indicate Dobson and optical filter (generally M-83) observations respectively. Locations are shown only for stations that have at least one complete year of observations. An underline designates stations where observations were made in 1980. There are only 10 stations in the global network missing fewer than six months of data for the entire period 1958-1980. But about 60 stations have observations for over 10 years, and there are present plans to augment the international network with additional stations over sparse data regions (WMO, 1981). Most of the ground stations for which observations are available are located in midlatitudes of the Northern Hemisphere and, except for a few island stations, over extended land areas. The data used for the analysis shown in Figure 1-1 (*A*) are not homogeneous in time and obviously are not uniformly distributed over the globe. Isolines of total ozone are drawn as dashed lines in areas where there are no ground-based observations. The ozone distribution shown in Figure 1-1 (A) is representative of the long-term annual average over a large part of the globe and is generally consistent with the shorter-period annual average distribution derived from satellite observations as discussed next.

The broad features of the global ozone distribution are reasonably well established (see, for instance, London et al., 1967, Shimizu, 1971). For an annual average, total ozone is a minimum of slightly less than 260 m-atm cm at equatorial latitudes and increases poleward in both hemispheres to a maximum of about 400 m-atm cm at subpolar latitudes. The high-latitude maximum is a result of transport of ozone from the region of primary production in the equatorial middle and upper stratosphere to the lower stratosphere in polar regions, where it has a relatively long

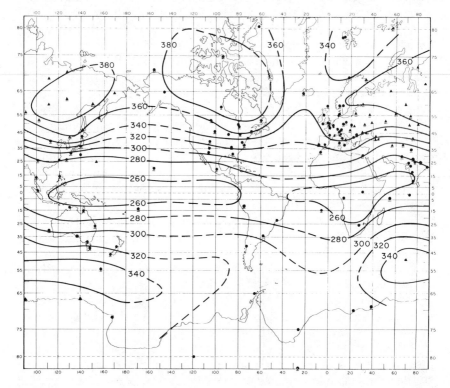

Figure 1-1 (A). Long-term annual average distribution of total ozone from ground-based Dobson (●) and optical filter (▲) observations (1958–1980). Underlined stations indicate observations in 1980. Dashed lines are for regions where there are no data. Units: m-atm cm.

photochemical relaxation time. The ozone maximum is higher and occurs at more poleward latitudes in the Northern than Southern Hemisphere. Northern Hemisphere maxima are found over Siberia and Northern Canada. Because of the scarcity of observations, it is not possible to be certain of the location of regions of ozone maxima in the Southern Hemisphere from ground-based measurements alone. However, the existence of a weak center of high ozone southwest of Australia in the South Indian Ocean is confirmed by satellite observations (e.g., Ghazi, 1980).

The most extensive set of satellite total ozone measurements available at present is that taken with the Nimbus-4 BUV instrument (April 1970 to May 1977). However, for reasons already discussed, the observation density decreased after the first two years of observations, and the instrument degradation resulted in a Dobson/BUV total ozone difference of about 10 Dobson units at the time of satellite launch, which increased

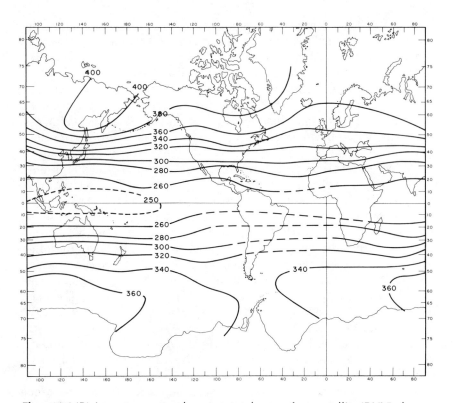

Figure 1-1 *(B)*. Long-term annual average total ozone from satellite (BUV) observations (April 1970–March 1977). Heavy dashed lines are in the region of no satellite data. Units: m-atm cm.

with time. The average annual global distribution of total ozone is shown in Figure 1-1 (B) as derived from the BUV observations. Although there are time and geographic gaps in the measurement distribution (there are, for instance, no observations poleward of about 70° during the winter or in the region of the so-called South Atlantic Anomaly, 10°N-40°S, 100°W-20°E), the total number of observations and the areas covered far exceed those made at ground base stations. The ozone patterns shown in the two diagrams are generally similar although there are differences in detail. The equatorial minimum given by the satellite observations is somewhat lower than the ground-based measurements, and the polar maximum is somewhat higher. The high latitude difference in ozone amount is most noticeable in the Southern Hemisphere where, as indicated, there are very few ground-based measurements. Two years of SBUV measurements (November 1978-October 1980) give slightly lower equatorial values than those shown for the BUV data, but the general pattern is the same.

As pointed out earlier, measurements made with broad-band filter instruments, particularly those made with the M-83 prior to 1972, tend to overestimate the total ozone amounts for low sun observations (at high latitudes during the winter). However, satellite observations for Eastern Siberia show an even stronger ozone maximum than the values given by the ground based measurements. Despite the similarities in the overall patterns as derived from the ground-based and satellite measurements, there may be significant differences for individual monthly distributions (Miller et al., 1981).

The long-term average latitude/season variation is shown in Figure 1-2 (A) for the station data and in Figure 1-2 (B) for the satellite-derived BUV data (April 1970 to March 1977). In each case the values were extrapolated poleward of 70° during the winter season since adequate observations were not available. The broad features of the total ozone distribution as derived from the two different methods are the same despite differences in time periods and geographical extent of the data coverage. There is an obvious strong ozone dependence on latitude and season, as was first shown by Dobson and colleagues over fifty years ago (Dobson, Harrison, and Lawrence, 1929).

The equatorial minimum of about 240-260 m-atm cm has a slight seasonal variation with lower values occurring during November-January. The latitudinal increase is strongest during the spring for each hemisphere, and total ozone reaches an annual maximum of about 460 m-atm cm in March-April in the Northern Hemisphere at about 75°N. The ozone maximum in the Southern Hemisphere, about 400 m-atm cm, occurs during September-October at about 60°S but is delayed by 1-2 months toward the south pole. In each case there also seems to be a slight ozone

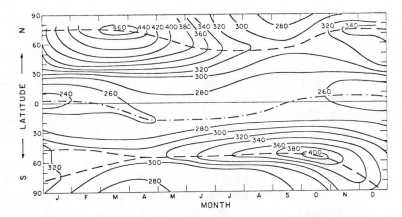

Figure 1-2 *(A)*. Long-term average latitude/season distribution of total ozone from ground-based data (1958–1980). Dashed lines show maximum ozone; dashed-dot line shows minimum ozone. Units: m-atm cm.

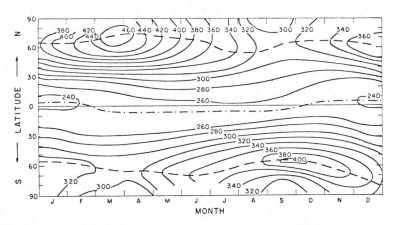

Figure 1-2 *(B)*. The same as Figure 1-2 *(A)* for satellite (BUV) data (April 1970–March 1977). Units: m-atm cm.

decrease towards the poles. Thus, the maximum is stronger and at a more poleward latitude in the Northern as compared to the Southern Hemisphere. During the fall season, the subpolar maxima are considerably weaker and, in the Northern Hemisphere, displaced equatorward. The seasonal and latitudinal variation of the region of maximum ozone is very clearly tied to seasonal and hemispheric differences in the stratospheric circulation at levels of 25-100 mb, a result of the variations in the equator to pole temperature gradient and the different topographical features of the two hemispheres (see, for instance, Cunnold et al., 1975; Murgatroyd, 1980; and the extended discussion in Mahlman et al., 1981).

The average annual variation is shown for each hemisphere in Figure 1-3. The solid and dashed curves are for ground-based measurements for the Northern and Southern Hemispheres respectively. The crosses and circles are for comparative satellite measurements, but for the period April 1970 to March 1977. In each case the winter data were approximated at latitudes poleward of 70° to account for the absence of observations in those regions.

There is a pronounced annual variation in each hemisphere, but with opposite phases and a larger amplitude in the Northern Hemisphere. The annual average difference between the two hemispheres, about 3%, is largely due to the difference in springtime maxima—an effect of the stronger poleward ozone transport in the Northern Hemisphere. The satellite-derived annual amount for each hemisphere is about 3 m-atm cm less than that determined from the ground-observed data. This amount is somewhat less than the differences noted when concurrent measurements between the different observing systems were intercompared (see, for instance, Miller et al., 1981). The results of satellite observations at equatorial latitudes, where there are very few ground stations, are lower

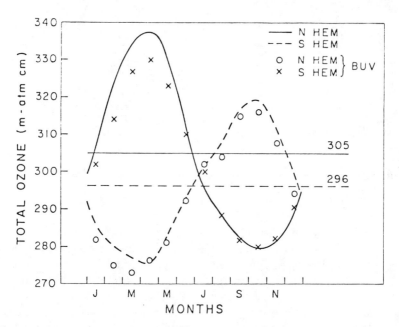

Figure 1-3. Average annual variation of total ozone for each hemisphere from station (1958–1980) and satellite (BUV) (April 1970–March 1977) data.

than those shown for the ground-based analysis. However, it is not clear why the differences are largest in January, February, and March for both hemispheres, and April for the Northern Hemisphere. Averaging of the two hemispheric curves in Figure 1-3 gives a global maximum during April with a secondary maximum in September/October, consistent with the results of the analysis of satellite observations discussed by Tolson (1981).

Periodic Variations. As indicated earlier, ozone variations have a number of periodic components, the most significant of which are annual, semiannual, quasi-biennial (at least in the tropics), and possibly those associated with solar activity. No periods shorter than a few months have been found from analysis of zonally averaged daily satellite data (Hilsenrath and Schlesinger, 1981). The annual and semiannual variations will be discussed next. Other observed periodic variations will be described in a later section.

There is a pronounced annual variation of total ozone that is strongly latitude dependent. A Fourier decomposition of the principal components of this fluctuation has been discussed by, for instance, Wilcox, Nastrom, and Belmont (1977) and London (1978a) based on long term records of station measurements, and by Tolson (1981) and Hilsenrath and Schlesinger (1981) using a large portion of the satellite BUV data.

For each latitude (l) the ozone variation is given by

$$X_l = \overline{X}_l + \sum_n C_{l,n} \cos (2\pi n \, [t - t_{l,n}]/P), \qquad (1\text{-}5)$$

where $\overline{X}_l$ is the latitude average ozone over the fundamental period P (assumed here to be 12 months), and $C_{l,n}$ is the amplitude, $t_{l,n}$ the phase of maximum for the nth harmonic. The latitude variation of the long term annual mean, the amplitude and phase of the 1st (annual) and amplitude of the 2nd (semiannual) harmonic are shown in Figure 1-4. Comparative values derived from the BUV observations are indicated by open circles. The satellite data were computed from the 7-year zonally averaged monthly mean values for the period April 1970–March 1977 and are very close to the results discussed by Hilsrenrath and Schlesinger (1981) who used latitudinally averaged daily means for essentially the same period, and to those discussed by Tolson (1981) derived from a spherical harmonic analysis of a latitude/longitude grid of monthly mean BUV data.

The mean annual ozone distribution shows a subpolar maximum in both hemispheres, but the Northern Hemisphere maximum is larger and farther poleward than that in the Southern Hemisphere. It should be noted that the interhemispheric asymmetry occurs only poleward of about 40° where there are significant differences in poleward stratospheric ozone transport. As indicated earlier, the satellite results give less ozone at equatorial latitudes and more ozone at polar latitudes, particularly in the Southern Hemisphere. This may be due to latitudinal differences in the bias between BUV and ground measurements as discussed by Keating, Lake, and Nicholson (1981).

The annual variation is dominant at all latitudes, particularly poleward of about 25° and at its maximum is about twice as large in the Northern as it is in the Southern Hemisphere. There is a small semiannual component to the seasonal ozone variation, whose amplitude is almost constant between 40°N and 40°S and increases somewhat at polar latitudes of each hemisphere. The variance of the observed distribution accounted for by the 1st (annual) harmonic is quite high (more than 90% at all latitudes except the southern tropics 0-20°S) and south of 60°S (London, 1978a; Hilsenrath and Schlesinger, 1981). The semiannual component in polar regions results from the seasonal asymmetry of the ozone variation. In the tropics, however, the small semiannual variation probably is due to the alternating influence of the Northern and Southern Hemisphere annual variation, since there is a tendency for the phase of the semiannual variation in the tropics to reflect the dominating annual oscillation.

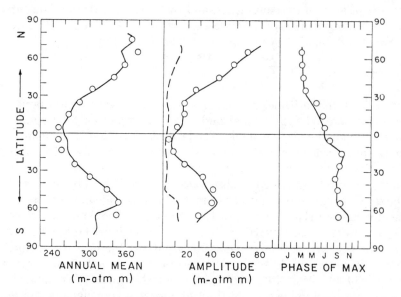

Figure 1-4. Latitude variation of long term mean, amplitude, and phase of the annual variation (solid line) and the amplitude of the semiannual variation (dashed line) derived from groundbased observations. The annual mean, amplitude and phase of the annual variation as derived from 7 years of BUV observations are shown as open circles.

The phase of the annual variation shows a maximum in mid-March at northern subpolar latitudes, is delayed to early April at midlatitudes, and further delayed to early June near the equator. An abrupt phase shift occurs at the equator with an indicated ozone maximum in spring at midlatitudes and late spring at subpolar latitudes of the Southern Hemisphere (see also Figures 1-2A and B). The mid-March maximum results from the transport of ozone towards higher latitudes in the Northern Hemisphere, where its photochemical relaxation time is relatively long, by the large-amplitude stationary waves in the stratosphere during the winter. In spring, when the latitudinal temperature gradient decreases, the amplitude of the stationary waves is much smaller and the poleward ozone transport is weakened. The slight phase retardation at high latitudes in the Southern Hemisphere is the result of the delayed breakdown of the Southern Hemisphere polar vortex, which allows continued ozone transport southward by the high frequency transient waves at subpolar latitudes.

Longitude Variations. In addition to the strong ozone dependence on latitude, there is a longitude variation that is most apparent in the Northern Hemisphere and is best developed during the late winter and early spring (see, for instance, London, 1963; London, Frederick, and Anderson, 1977). The variation is such that high ozone amounts are generally found at longitudes corresponding to upper-level trough positions and vice versa (e.g., Miller et al., 1979). This relationship is much more pronounced during winter/spring when the circulation systems in the lower and middle stratosphere are closely linked and affected by surface topography. An analogous association is found between total ozone and the temperature distribution in the lower stratosphere during the spring (i.e., high ozone in regions of low temperature at 100 mb) (WMO, 1982).

Harmonic components of the observed longitude variation of ozone have been computed by Wilcox et al. (1977) from a limited set of station data, and by Tolson (1981) from 7 years of global satellite BUV data. The results discussed by Tolson indicate that wave number one is dominant at high latitudes during the winter in the Northern Hemisphere and the spring in the Southern Hemisphere. Wave numbers two and three are significant only at mid and subpolar latitudes in the Northern Hemisphere during the winter and spring. In summer and early fall the stratospheric latitudinal temperature gradient is reversed and the stratospheric winds are strongly zonal (from the east) and only minimally related to topographically induced standing waves. Thus, longitude variations in ozone during the summer are small.

An example of the longitude difference observed at two stations (Sapporo and Arosa) at approximately the same latitude has been discussed by Rangarajan and Mani (1981). The long-term (1958-1980) average monthly mean total ozone amounts for Sapporo (43N, 141E) and Arosa (47N, 10E) are shown in Figure 1-5. Although Arosa is at a slightly higher latitude than Sapporo, both maximum and minimum at Sapporo are higher and appear earlier than those at Arosa. The difference in maxima is greater by a factor of three as compared to the minima difference; the annual range at Sapporo is 50% larger than at Arosa. These differences arise from the pronounced N-S mean trough during the winter at 50-100 mb over Sapporo but not over western Europe. As a result, northwesterly flow from eastern Siberia accompanied by mean subsidence in the lower stratosphere both contribute to the large ozone increase over Sapporo during the winter. Evidence for a similar longitude variation of total ozone in the Southern Hemisphere has been presented by Kulkarni and Garnham (1970).

The global distribution of the annual range (i.e., the difference between the average highest and lowest mean monthly station values) is shown in Figure 1-6. The annual range is a minimum in the tropics (<25 m-atm cm) and a maximum in polar regions (>150 m-atm cm). The range is larger by a factor of two in the Northern Hemisphere poleward of about 20° and

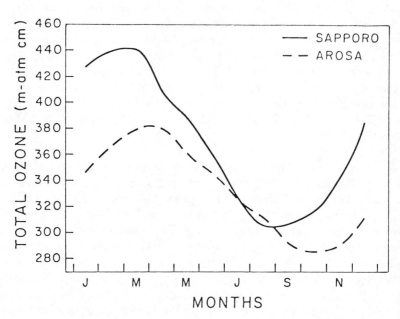

Figure 1-5. Comparison of the long term annual variation of total ozone at Sapporo (43N-141E) and Arosa (47N-10E) for the period 1958-1980.

has an obvious longitude dependence such that the maximum range occurs in regions of maximum winter/spring ozone. A similar analysis for 15 years of Northern Hemisphere ground-based measurements is summarized by Wilcox et al. (1977).

Quasi-biennial Oscillation. Soon after the discovery of a near 2-year periodic reversal of the zonal wind in the tropical stratosphere (Reed et al., 1961), a 24-month cycle was detected by Funk and Garnham (1962) from analysis of total ozone observations at Aspendale and Brisbane. Since that time, the ozone quasi-biennial oscillation (QBO) in the tropics and elsewhere has been studied using both station and satellite data (e.g., Ramanathan, 1963; Angell and Korshover, 1973; Hilsenrath and Schlesinger, 1981; Tolson, 1981; Oltmans and London, 1982; Zerefos, 1983).

Variations of total ozone are highly correlated with ozone variations in the lower and middle stratosphere where the ozone photochemical relaxa-

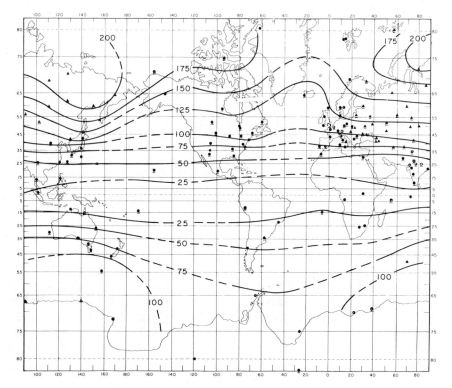

Figure 1-6. The global distribution of mean annual range of total ozone determined from ground-based observations (1958–1980). Units: m-atm cm.

tion time varies from one to several months. It is therefore to be expected that transport processes play a major role in producing the ozone oscillations. When the observed variations in the tropics are filtered for annual and semiannual periodicities and long term trends, the residual variation shows a strong relationship to the QBO of the zonal tropical wind in the stratosphere such that maximum ozone is associated with strong west winds. Observations also indicate that at the equator the QBO in ozone is positively correlated with the QBO in temperature at 50 mb (i.e., maximum ozone occurs with maximum temperature [e.g., Zerefos, 1983]). This latter result is consistent with vertical motion patterns at the equator postulated by Reed (1964) as part of his explanation for the tropical wind QBO. In the tropics, the average period of the ozone QBO is approximately 27 months and is closely in phase with the 30 mb zonal winds. Its amplitude is about 6-8 m-atm cm, approximately equal to the amplitude of the annual ozone variation. The amplitude seems to be constant, or increases slightly, with latitude in the Northern Hemisphere to a possible maximum at 25°N, decreases to almost zero between 40-50°N, and increases again at subpolar latitudes. In the Southern Hemisphere there is a QBO amplitude minimum at 10-15°S and a maximum at about 40°S (Hilsenrath and Schlesinger, 1981; Tolson, 1981). This pattern of the asymmetric amplitude variation parallels that of the cross-spectral coherence between the total ozone oscillations and the 50 mb tropical wind QBO (Oltmans and London, 1982; Zerefos, 1983).

At latitudes where the ozone QBO amplitude is a minimum, the dominant period of the oscillation is about 24 months. This lower period, as compared to that at the equator, might be a reflection of the interacting influences and mutual modifications of the tropical wind oscillation and the seasonal variation of the midlatitude stratospheric circulation (e.g., Tucker, 1979; Holton and Tan, 1980). Poleward of the QBO amplitude minimum there seems to be an abrupt phase retardation of about 13-14 months with respect to the equatorial QBO. There is as yet no satisfactory explanation for the latitudinal variation of the amplitude and phase of the ozone QBO.

Aperiodic Changes. In addition to the observed periodic variations discussed above, there is a residual nonperiodic variance, in both daily and mean monthly data, that is a function of latitude and season and varies with time. The variance increases with latitude and is generally higher in the Northern than in the Southern Hemisphere. The standard deviation of the mean monthly residual total ozone also shows a longitude dependence and can vary by as much as 50% over a few years (London, 1980). The average normalized standard deviation of the daily values can be of the order of 12% at high latitudes during the winter months and is about 2% in the tropics during all months of the year. The largest contribution to these variations comes from the stratospheric layer 15-30 km, where large disturbances to the mean winter circulation are directly associated with local synoptic scale perturbations, particularly at middle and subpolar latitudes (e.g., Dobson, 1973). The latitudinal and seasonal pattern of the ozone variance is very similar to analogous variance patterns of meteorological variables such as pressure, temperature, and winds in the lower and middle stratosphere (e.g., Van Loon, Labitzke, and Jenne, 1974; Van Loon and Jenne, 1974).

In addition to the day-to-day and month-to-month ozone variations, there are interannual changes that are for the most part unexplained. These long term fluctuations are probably related to large-scale changes in the tropospheric and stratospheric circulation characteristics, but they may also be influenced either directly or indirectly by anomalous solar variability, anthropogenic effects that perturb the stratospheric composition, or as yet undisclosed causes. Part of the difficulty of documenting extended period ozone changes is that of extracting the relatively small signal of the long-term variation from the much larger short period fluctuations, which include significant geographical differences. Satellite observations have greatly increased the space and time density of ozone measurements. But the total period for which these observations are available is at present quite limited, and the instrumental uncertainties of the measurements are of the order of the changes that have been inferred from other observational methods or theoretical models.

The long measurement records from ground-based instruments have been used frequently in an attempt to determine the existence of some pattern of extended changes. However, the ground-based network is not homogeneous in space or time. In addition, there have also been differences noted in the accuracy and precision among the various instruments used in the global network. Nevertheless, the long-term patterns, both periodic components and short-period variances of ozone as determined from the ground-based data, are quite consistent with the limited results derived from the satellite measurements. Sophisticated time-series models are currently being applied in an attempt to extricate information of long-period trends from the available space and time data ensemble derived from both station and satellite measurements (e.g. Reinsel et al., 1981; WMO, 1982).

From an analysis of the mean monthly Dobson data at seven stations, Komhyr et al. (1971) inferred that there was an increase of ozone during the period 1960-1970 of about 5% per decade. This was generally confirmed by London and Kelley (1974) in a study of 13 years of monthly mean ozone data derived from the global ozone network. However, it was noted that the change, though strongly positive during this period in the

Northern Hemisphere, was essentially zero in the Southern Hemisphere. The results of subsequent studies (see, for instance, Angell and Korshover, 1976; 1978; Zerefos, Repapis, and Jenne, 1982), using station and satellite data, indicate that for a few years after 1970 there probably was a small decrease in the Northern Hemisphere but no significant systematic change in the Southern Hemisphere.

The interannual total ozone variation, as determined from measurements at ground stations during the period 1958-1980, is shown in Figure 1-7. The curves are based on mean values for 10° latitude belts and averaged over the latitude sectors 0°-60° (top curves) and 30°-50° (bottom curves) for each hemisphere. Each of the top curves covers over 85 percent of the area of the indicated hemisphere and is reasonably representative of the hemispheric average. In the Southern Hemisphere there were many years of incomplete station data, particularly before 1964, and the values shown for that time have been estimated from smoothed data. The bottom curves cover the latitude span where most of the station observations are made and should, at least in the Northern

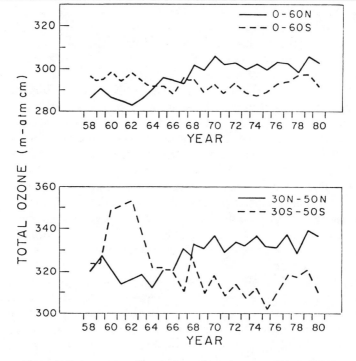

Figure 1-7. Interannual variation of total ozone (1958–1980).

Hemisphere, give a reasonably good picture of the year-to-year variation at those latitudes.

As already mentioned, total ozone increased in the Northern Hemisphere, particularly after 1962, to a maximum in 1970 and decreased slightly, if at all, during the next 10 years. In middle latitudes of both hemispheres the long period changes over the 23 years shown are apparently negatively correlated. But it should be noted that the Southern Hemisphere values are based on data from at most three stations in the midlatitude belt prior to 1965. The interannual variations shown for the period 1970-1977 are in reasonable agreement with those derived from the BUV satellite measurements (Zerefos, Repapis, and Jenne, 1982).

An increase of total ozone during the time of seasonal ozone minimum (September or October) was noted by Vernazza and Foukal (1980) from ground-based measurements in the Northern Hemisphere for the period 1957-1975. There was, apparently, no such long term change during the time of ozone maximum (February, March, or April). However, this increase of the ozone minimum occurred largely prior to 1970, and analysis of the total ozone data for the 10-year period 1970-1980 does not show a seasonal difference in long-term trends.

Computed long-term ozone trends are significantly influenced by the time-averaging interval and can be affected by strong short-term local perturbations (WMO, 1982). For example, the average observed large ozone increase followed by an equally large decrease in the early 1940s over Western Europe was most likely associated with the major winter and spring circulation perturbations affecting Western Europe in 1940 and 1941. Variations over extended periods of time may also be different for different stations and different geographical regions (e.g., London, 1978*b*; Angell and Korshover, 1981). Thus, hemispheric or worldwide averages of long-term ozone variations could very well mask the physical basis for these changes where the changes have a strong geographic pattern as occurs, for instance, with other meteorological variables.

The Vertical Distribution

Umkehr and optical balloon measurements in the 1930s provided a general picture of the vertical ozone distribution in the upper troposphere and stratosphere. However, only during the past 15-20 years has sufficient information become available to describe the height/latitude/time variations of the ozone concentration in the stratosphere and, with the advent of both nadir and limb viewing satellite observations, its geographic distribution. The current routine global measurement programs involve Umkehr, ozonesonde and satellite observations. Rocket measurements have contributed significantly to the information bank by showing the vertical ozone distribution in the stratosphere and mesosphere. They have also provided an important link between ground-based, balloon and different satellite measurement systems for coordination and calibration purposes. However, there are no long-range plans at present for an international program of regular rocketsonde observations.

The longest set of vertical profile data is that provided by Umkehr observations that have been made routinely at a few places since 1956. Some observations have been reported over the years from about 60 different places, and at present there are almost 25 locations where an average of 10-15 observations are made per month. A few regular ozonesonde observations were taken in the mid-1960s, and an extended program of measurements using the Regener chemiluminescent sonde was conducted over the North American continent during the period 1963-1965 (Hering and Borden, 1967). However, because of difficulties in quality control, as discussed earlier, the Regener sondes were not found suitable for continued routine stratospheric measurements. The present global ozonesonde

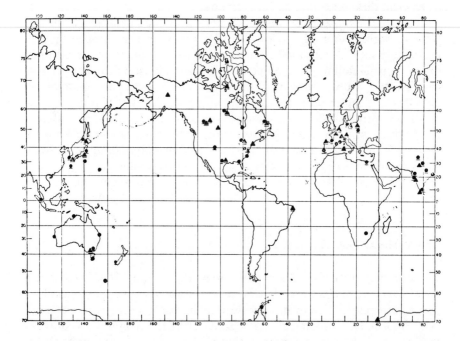

Figure 1-8. Geographic distribution of locations where ground-based (Umkehr) and ozonesonde ozone profile measurements are made. Underline indicates observations in 1980.

network, based primarily on different electrochemical systems, was started in 1966. This network currently involves measurements from about 20 stations, most of which are in midlatitudes of the Northern Hemisphere. In general, ozonesonde observations are taken only about 3-4 times per month, although as many as 10-12 observations per month are taken at a few places. The distribution of locations where Umkehr and ozonesonde observations have been made is shown in Figure 1-8. Six stations in the global network presently make both types of measurements. Underlined locations indicate places in current operation.

Stratospheric ozone concentrations derived from ozonesonde and Umkehr techniques generally give consistently different absolute values. Comparison of the long-term monthly ozone average partial pressure in different layers over central Switzerland, as derived from Umkehr and ozonesonde observations, has been discussed by Dütsch and Ling (1973); (see also WMO, 1982; London and Angell, 1982). A similar comparison for long-term averaged annual profiles over Sapporo is shown in Figure 1-9. In the troposphere and lower stratosphere, the two sets of measurements are in good agreement. But in the region of ozone maximum, 25-80 mb, the Umkehr-derived values are appreciably lower than those measured in situ by the ozonesonde system. This results primarily from the smoothing process that is intrinsic to the Umkehr evaluation technique. Above the layer of maximum ozone, this process produces Umkehr values slightly higher than those given by the ozonesonde observations. This may in part stem from the instrumental difficulties found with the balloon measurements above 10 mb.

The studies over Switzerland referred to above indicated that the seasonal variation derived by both methods are in good agreement with each other. They show a late winter/spring maximum in the midtroposphere, a strong winter/spring maximum in the lower and midstratosphere and a weak summer maximum at about 10 mb. The Umkehr measurements for the top layer (1-2 mb) show the highest amount during the winter. Umkehr observations are generally more reliable above 30 km than in the region of the ozone maximum. It should be noted, however, that Umkehr observations are highly biased towards weather patterns involving conditions of clear or only partly cloudy skies. Thus, many regions that have pronounced seasonal variations of cloudiness such as summer monsoon conditions are subject to strong biases in the reported Umkehr measurements. Comparisons have also been made between cosynchronous Umkehr and BUV measurements (e.g., DeLuisi and Nimira, 1978; WMO, 1981). The vertical profiles given by the two systems appear to be in mutual accord above about 10 mb with the average Umkehr amounts being slightly lower than those derived from the satellite.

Average Vertical Distribution. The observed ozone concentration is nearly constant in the troposphere, but it increases with height in the lower stratosphere to a maximum that depends on latitude and season. Above the level of maximum, the concentration decreases almost exponentially with height through the middle and upper stratosphere

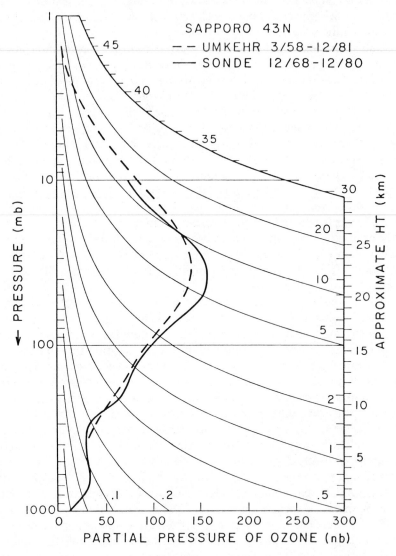

Figure 1-9. Comparison of Umkehr and ozonesonde ozone profile measurements over Sapporo.

and lower mesosphere. Recent satellite observations have confirmed the results of earlier rocket measurements showing that there is a slight secondary maximum in ozone concentration a few kilometers above the mesopause (Thomas et al., 1983*a*).

In general, the lower the height of the stratospheric maximum concentration, the larger is the ozone amount at that height. Maximum ozone is found at about 25-27 km in the tropics and at about 18 km in polar regions. The ozone partial pressure at the height of maximum is about 140 nb in the tropics and about 220 nb at polar latitudes during the spring. The ozone mixing ratio (ppmv), however, has its maximum in the tropics at a height of about 33-35 km. The vertical distribution of the ozone partial pressure is generally related to the total ozone amount so that low maximum heights and large maximum concentrations are associated with large total amounts. The climatology of the vertical ozone distribution and its relation of the meteorology of the lower stratosphere has been discussed by Dütsch (1978, 1980).

An example of the typical latitude variation of the vertical ozone distribution can be seen from the mean annual ozone profiles for Resolute (75N) and Wallops Island (38N), shown in Figure 1-10, as determined from ozonesonde observations. The mean annual total ozone at Resolute and Wallops Island is 391 m-atm cm and 329 m-atm cm respectively. The height of maximum concentration at Resolute is about 5-6 km lower than at Wallops Island, and the ozone concentration is much greater at all levels from 40 to 400 mb (7-22 km). The latitudinal gradient is reversed in the lower troposphere and in the middle stratosphere, above about 30 mb. The larger lower tropospheric value at Wallops Island results from the stronger vertical mixing at midlatitudes, particularly during the spring and summer months. Above 30 mb, the reversed latitudinal gradient is probably caused by photochemical effects above the layer of maximum ozone. Thus, the larger difference in total amount between these two stations is clearly a result of the higher ozone concentration in the lower and middle stratosphere.

Analyses of ozone profile data have shown that the correlation between total ozone and the ozone concentration at different heights is usually a maximum at about 50-100 mb for midlatitude stations. Approximately 50% of the monthly variance in total ozone can be accounted for by the variance in the ozone concentration at these levels (e.g., Dütsch, 1980; London and Angell, 1982). Since there are many more total ozone than vertical profile observations, these correlative features have been used in estimating stratospheric ozone transport based on observed 100 mb wind patterns (e.g., Newell 1961, 1964). It is now possible, of course, with the availability of satellite measurements, to compute such transport directly (e.g., Gille et al., 1981).

There is a significant seasonal variation of the vertical distribution, which as for total ozone is also strongly latitude dependent. The long-term average profiles for four individual months for Resolute (75°N) (Fig. 1-11A) and Kagoshima (31°N) (Fig. 1-11B) represent polar and subtropical variations. At Resolute, the height of the ozone maximum is about

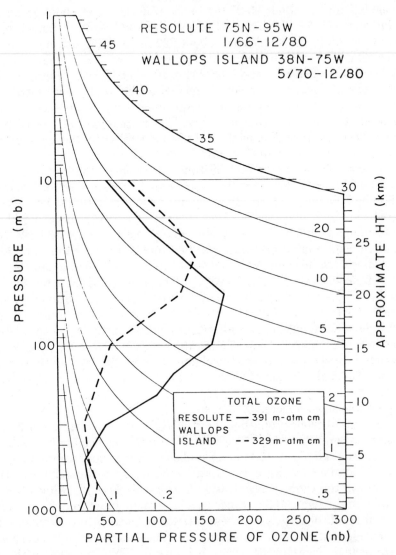

Figure 1-10. Mean annual vertical ozone profile at Resolute (75N) and Wallops Island (38N) from ozonesonde measurements.

15-20 km during winter and spring (January and April) and about 21 km during summer and fall (July and October). The ozone concentration is distinctly greater in the winter and spring at all levels between 10 and 500 mb than in summer and fall. Even at 30 km (10 mb) there continues to be a small winter/spring maximum that persists until the high-latitude winter circulation pattern breaks down in late spring. As will be shown later, a winter maximum in polar regions is also found in the upper stratosphere as a result of the temperature-dependent photochemistry at those levels. At Resolute there does not seem to be any significant seasonal variation below 500 mb.

At Kagoshima the ozone peak is found at about 25 km (25 mb) during all months of the year. Although there is a slight spring maximum at all levels below about 22 km (a bit more pronounced in the layer 100-200 mb), the seasonal variation is otherwise typically small. The distributions shown for Kagoshima are very similar to other subtropical observations (e.g., Craig, Deluisi, and Stretrer, 1967).

The latitudinal variation of the vertical distribution has been generalized by Dütsch (1978, 1980) mainly on the basis of ozonesonde measurements made during the period 1963-1974. Since most of the balloon observations were taken at midlatitude stations in the Northern Hemisphere, Umkehr and one year of BUV measurements were used to broaden the data base of the analysis. The average ozone profiles for 20° latitude belts in the Northern Hemisphere are given in Figure 1-12 (Dütsch, 1980) for April and October.

During both months the ozone partial pressure is approximately constant at each latitude up to the base of the stratosphere and then increases sharply with height in the lower stratosphere. The height of the ozone maximum decreases with latitude from about 27 km in the tropics to about 18-20 km at polar latitudes. The latitudinal increase in total ozone, as shown in Figures 1-2 (A) and (B), is the result of the large latitudinal difference in low stratospheric ozone. The strong poleward increase of the ozone concentration between 30 mb and 400 mb, while present in both months, is much more pronounced in April. Above 30 mb, to about 5 mb, there is a weaker reversed gradient (i.e., an increase toward the equator) that is slightly smaller in April than October. This reversed gradient reflects the increasing solar control on the photochemical influence on the ozone concentration in the middle stratosphere above the ozone peak. Seasonal variations are significant only at middle and high latitudes, and are apparent in the concentration at the ozone peak rather than the location of the peak. In the tropics there is no appreciable latitude difference at any level.

Latitudinal variations of total ozone result from a balance between the strong increase poleward in the lower stratosphere and the weaker decrease in the middle stratosphere. The seasonal difference in the observed total ozone latitude gradient is thus produced by the large increase in the ozone concentration in April. In October, the lower stratospheric poleward increase is smaller and the equatorward increase in the middle stratosphere is somewhat larger than in April. This combination explains the

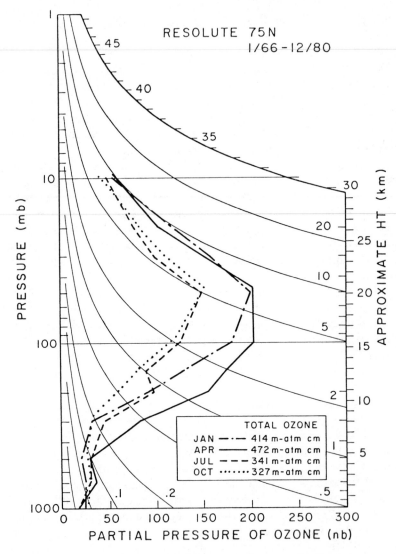

Figure 1-11 *(A)*. Seasonal variation of the vertical ozone distribution at a polar station (Resolute 75N).

very strong latitude gradient of total ozone in April and weak gradient in October. The seasonal difference in the lower stratosphere is a result of the seasonal difference in poleward eddy transport of ozone in the lower transport. The springtime flux is much higher than during fall (e.g., Hering, 1966; Cunnold, Alyea, and Prinn, 1980).

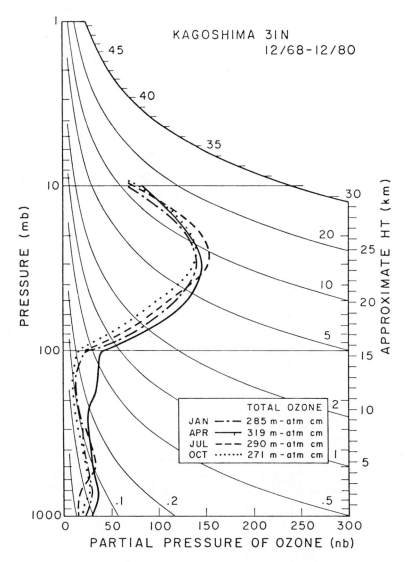

Figure 1-11 *(B)*. Seasonal variation of the vertical ozone distribution at a subtropical station (Kagoshima 31N).

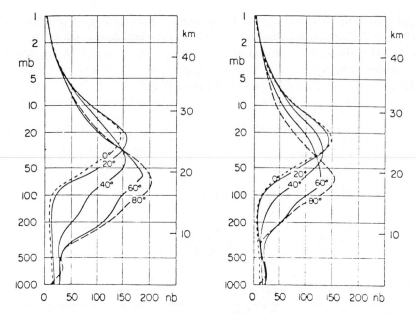

Figure 1-12. The average latitude variation of the vertical ozone distribution. *(From Dütsch, H. U., 1980, Vertical ozone distribution and tropospheric ozone, in Proceedings of the NATO Advanced Study Institute on Atmospheric Ozone: Its Variation and Human Influences, A. C. Aiken, ed., Rept. No. FAA-EE-80-20, U.S. Department of Transportation, Federal Aviation Administration, Washington, D.C., p. 19).*

Interhemispheric differences are largest at high latitudes in winter and spring when there is less ozone at south than north polar regions (Dütsch, 1978). The breakdown of the lower stratosphere (100 mb) Antarctic polar circulation in late spring (Knittel, 1976) results in a delayed maximum, as also occurs for total ozone.

As discussed earlier, total ozone variations, although largely latitude dependent, have significant longitude differences at mid and subpolar latitudes. These differences represent a response to the longitude variation of the ozone concentration in the upper troposphere and stratosphere, as is illustrated in Figure 1-13, where long-term average Umkehr observations for Sapporo and Arosa are compared. The curves are for January, the month of largest vertical profile difference between these two stations. There is practically no difference between the two distributions in July. The mean ozone concentration at Sapporo is distinctly larger than at Arosa in the interval 20-300 mb, where the ozone transport is most important.

Sapporo is located in a geographic region of a pronounced winter upper level trough, and horizontal advection from the northwest combined with persistent subsidence west of the trough result in a stronger ozone increase at Sapporo than at Arosa, where perturbations to the mean winter zonal flow pattern are not as extreme. During the summer (July) the midlatitude winds above 100 mb are nearly zonally symmetric, and

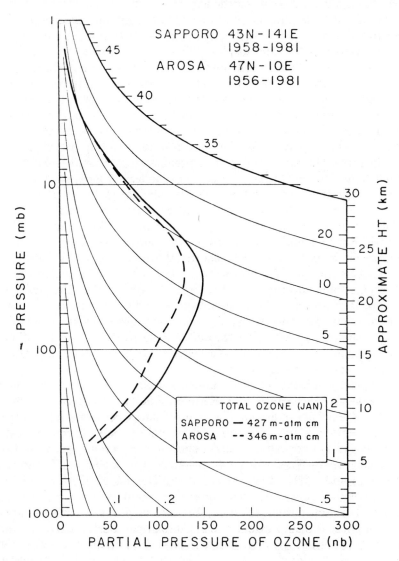

Figure 1-13. Mean vertical profile curves for Sapporo and Arosa taken from Umkehr data for January.

the longitude ozone variations are minimal. As shown by satellite observations (e.g. London, Frederick, and Anderson, 1977; Gille, 1980), the average longitude differences tend to decrease above about 10 mb, where the influence of photochemistry on the ozone changes becomes dominant. Even in the middle and upper stratosphere, however, day-to-day variations related to high frequency temperature changes are found particularly at high latitudes during the winter.

As a result of satellite observations made during the past 15 years, it is now possible to provide a global distribution of the ozone concentration in the stratosphere and mesosphere. The various satellite systems listed in Table 1-2 use different measurement techniques and have been operative for different lengths of time and at different time periods. Nevertheless, the results derived from these systems are generally mutually consistent and can be used to portray the general features of the upper air ozone distribution. There are sufficient data available also to determine the pattern of the annual and semiannual variation as a function of pressure and latitude. Current satellite measurements can provide information about high frequency (days to weeks) ozone oscillations, such as those related to synoptic variations and 27-day solar variability. BUV observations measured the strong sudden ozone decrease at high latitudes above 4 mb that resulted from the enhanced solar proton flux in August 1972 (Heath, Krueger, and Crutzen, 1977). Measurements using the airglow instrument on the SME satellite reported the mesospheric ozone depletion, reaching as high as 70% over some parts of the Northern Hemisphere polar cap, following the July 1982 Solar Proton Event (Thomas et al., 1983*b*). Differences in absolute calibration and the relatively short time period of most data sets, however, preclude at present anything more than gross estimates of the QBO in the middle stratosphere or the detection of small changes such as might be associated with long-period solar cycle variations or anthropogenic causes.

The most extensive set of continuous satellite ozone measurements of stratospheric concentrations currently available is that derived from the Nimbus-4 BUV observations. The reported data cover a seven-year period for the interval 30 mb to 0.7 mb. Additional satellite observations that provide similar information, but for shorter time periods, are those from OGO-4 (London, Frederick, and Anderson, 1977), SBUV (Frederick et al., 1983) and SME (Rusch et al., 1983; Thomas et al., 1983*a*). Since Umkehr and ozonesonde observations are available over much longer periods than any satellite system, it is also helpful to use these data to normalize the different satellite sets. Comparison studies of results from the various observing systems (DeLuisi, Mateer, and Heath, 1979; Fleig et al., 1981; Frederick et al., 1983; Bhartia et al., 1984) indicate that whereas the seasonal variations determined for each level from the different observing systems are quite similar, the absolute values of the concentration distributions are somewhat different. In general, the BUV and SBUV amounts are larger than those given from Umkehr measurements at 40-45 km and lower than the Umkehr values at 25-30 km. This latter difference could account for the reported BUV/SBUV total ozone values being lower than those given by the ground-based global network. In addition, there are latitude and seasonal differences between the two satellite measurement sets that, however, may indicate interannual variations of the ozone concentration.

The average pressure/latitude distribution of the ozone mixing ratio (ppmv) is shown in Figure 1-14 (*A*) (January) and Fig. 1-14 (*B*) (July). The analysis is based primarily on the BUV observations (30-0.7 mb), adjusted for a long-term linear decrease due presumably to an instrumental drift (Fleig et al., 1981) and on SME-UVS measurements for the interval 1.0-0.1 mb. The BUV data were adjusted slightly to be consistent with ozonesonde and Umkehr data above 50 mb. In the overlap region near 1 mb, the BUV and UVS data were generally in agreement within about 10%. Similar diagrams based mainly on ozonesonde and Umkehr data have been discussed by Dütsch (1978) for all four seasons. It can be seen from Figure 1-14 (*A*) and (*B*) that the ozone mixing ratio increases in the stratosphere to a maximum of slightly more than 10 ppmv in the tropics at a height of about 30-35 km and then decreases through the upper stratosphere and lower mesosphere. The midstratosphere ozone peak shifts with season and latitude, increasing in height poleward, particularly in the winter hemisphere. The ozone maximum occurs at 10-15°S in January and at 10-15°N in July, in parallel to the shift in the subpolar latitude. It is not clear whether or not the small increase in this maximum concentration in January is a result of the changing distance of the earth to the sun. In the upper stratosphere, up to about 55 km, there is a distinct winter maximum at high latitudes with a small seasonal variation (maximum in June-July) shown in the tropics. Such a variation in the tropical upper stratosphere and lower mesosphere was suggested by Frederick et al. (1980, 1981) on the basis of backscattered ultraviolet radiance measurements and by the SME-UVS observations. At 50 mb there is a mixing ratio maximum at 50-60° latitude in the winter hemisphere that results from poleward and downward transport at these latitudes from the high-level tropical source region as discussed earlier.

Periodic Variations in the Stratosphere. Analysis of the periodic variations of the ozone mixing ratio in the stratosphere (London, 1978*b*, 1983) shows that the amplitude of the annual variation is very small at the equator and at all levels increases poleward to a maximum at mid and subpolar latitudes. The annual variation at these latitudes is quite large in the lower stratosphere (15-20 km) and decreases to a minimum at about 25 km. Both satellite and Umkehr observations at subpolar latitudes indicate a strong annual fluctuation of the ozone mixing ratio, with an amplitude of about 30% of its mean annual value at that height. The large annual oscillation in the upper stratosphere is in response to the large seasonal temperature variation ($\sim$50C) at the high latitude stratopause where there is a strong temperature influence on the ozone photochemistry. In this region over 60% of the total variance of ozone is accounted for by its annual variation.

The phase of the annual variation changes significantly through the entire stratosphere. In the lower stratosphere, the phase is the same as that for total ozone, that is, the maximum ozone mixing ratio occurs

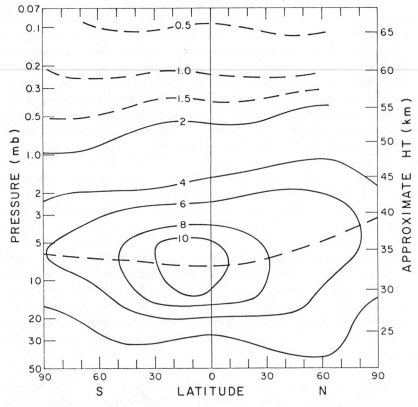

Figure 1-14 *(A).* Average ozone mixing ratio distribution (ppmv) from satellite observations (see text) in January.

during the spring for each hemisphere. The springtime maximum continues up to about 25 km. At levels of about 30-35 km the maximum is found
during the summer because of the dominant photochemical influence at
these heights. Above 35 km, where the ozone equilibrium values are
inversely temperature dependent through the odd oxygen recombination
system, the ozone mixing ratio has its annual peak during the winter.
Thus, there seem to be three distinct layers in the stratosphere and lower
mesosphere that respond to three dominant physical/chemical mechanisms for their annual variation: the lower stratosphere up to about 25-30
km, where the major influence involves stratospheric transport processes;
the region 30-35 km, where direct solar effects exert a dominant control
on the ozone variations; and the layers above 35 km (40-65 km), where
temperature influences on the ozone photochemistry are important.

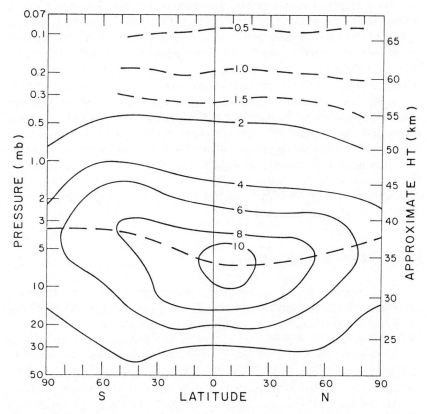

Figure 1-14 *(B)*. Average ozone mixing ratio distribution (ppmv) from satellite
observations (see text) in July.

At pressure levels in the lower stratosphere (25-50 mb), an ozone QBO is found in the tropics from Umkehr and BUV data (Oltmans and London, 1982). But this oscillation seems to be weak, at best, at lower pressures (above 10 mb) or at higher latitudes. A longer series of continuous satellite observations than is presently available, along with additional ozonesonde and Umkehr measurements in the tropics and subtropics, are needed to isolate the stratospheric ozone QBO from the other periodic and aperiodic variations.

CONCLUSION

The average global distribution of total ozone and the latitudinal dependence of the phase and amplitude of its seasonal variation as determined from the global ground-based observing network, and the various satellite measurements, are reasonably well known. There is information about the planetary patterns of the longitude variation, and details of the higher frequency structure have become available through recent satellite observations. The various measurements have shown that there is a close association between the planetary wave pattern of total ozone and the large-scale circulation patterns in the lower stratosphere.

During the past few years, satellite observations have particularly contributed to our understanding of the general features of the vertical ozone distribution and latitude variation in the stratosphere and lower mesosphere. Continued programs of stratospheric and mesospheric ozone measurements will contribute significantly to further understanding of the interaction of dynamics and ozone photochemistry in the middle atmosphere (25-35 km) and the nature of stratospheric responses to anomalous solar activity.

ACKNOWLEDGMENTS

I would like to acknowledge with appreciation the technical assistance in the preparation of this chapter provided by Xiu-de Ling, Derk Norton, Hannah Sable, and Mary Sable. My thanks are extended to Walter Komhyr for his critical review of the section on observational techniques. Some of the research summarized here was supported by research grant NSG 5153 from the National Aeronautics and Space Administration. Acknowledgment is also made to the National Center for Atmospheric Research, which is sponsored by the National Science Foundation, for computing time used in this research.

HISTORICAL BIBLIOGRAPHY

1879, Leeds, Albert R., Lines of discovery in the history of ozone with an index of its literature and an appendix upon the literature of peroxide of hydrogen, *N. Y. Acad. Sci. Ann.* 1:363-404. (See also *N. Y. Acad. Sci. Ann.*, 2nd mem., 3:137-155, 1883.)

1950, Fabry, Charles, *L'Ozone Atmosphérique,* Centre National de la Recherche Scientifique, Paris.

1959, American Meteorological Society, *Meteorological Abstracts and Bibliography* 10, nos. 5 and 6, May and June.

REFERENCES

Aimedieu, P., and J. Barat, 1978, Un ozonomètre stratosphérique à haute résolution spatiotemporelle, *Acad. Sci. (Paris) C. R.* série B, 287:337-340.

Aimedieu, P., J. Barat, C. Bernard, and J. Ohayon, 1981, Ozonomètres stratosphériques à chimiluminescence en phase gazeuse utilisant des ethyleniques comme reactifs, in *Proceedings of the Quadrennial International Ozone Symposium,* J. London, ed., National Center for Atmospheric Research, Boulder, Colo., 212-215.

Ainsworth, J. E., J. R. Hagemeyer, and E. I. Reed, 1981, Error in Dasibi flight measurements of atmospheric ozone due to instrument wall-loss, *Geophys. Res. Lett.* 8:1071-1072.

Anderson, G. P., C. A. Barth, F. Cayla, and J. London, 1969, Satellite observations of the vertical ozone distribution in the upper stratosphere, *Ann. Geophys.* 25:341-345.

Angell, J. K., and J. Korshover, 1973, Quasi-biennial and long-term fluctuations in total ozone, *Mon. Weather Rev.* 101:426-443.

Angell, J. K., and J. Korshover, 1976, Global analysis of recent total ozone fluctuations, *Mon. Weather Rev.* 104:63-75.

Angell, J. K., and J. Korshover, 1978, Global ozone variations: an update into 1976, *Mon. Weather Rev.* 106:725-737.

Angell, J. K., and J. Korshover, 1981, Update of ozone variations through 1979, in *Proceedings of the Quadrennial International Ozone Symposium,* J. London, ed., National Center for Atmospheric Research, Boulder, Colorado, 393-395.

Annals of the Astrophysical Observatory of the Smithsonian Institution, 1914, 1922; vols. 3 and 4.

Attmannspacher, W., and H. U. Dütsch, 1970, *International Ozonesonde Intercomparison at the Observatory Hohenpeissenberg,* 19 Jan.-5 Feb. 1970, Berichte des Deutschen Wetterdienstes Nr. 120, Meteorological Service, Hohenpeissenberg, F.R.G., 74p.

Attmannspacher, W., and H. U. Dütsch, 1981, *2nd International Ozonesonde Intercomparison at the Observatory Hohenpeissenberg,* 5-20 April 1978, Meteorological Service, Hohenpeissenberg, F.R.G., 65p.

Barbier, D., D. Chalonge, and E. Vigroux, 1942, Utilisation des éclipses de lune à l'étude de la haute atmosphère, *Acad. Sci. (Paris) C. R.* **214**:983-984.

Bass, Arnold M., and Richard J. Paur, 1981, Absorption cross-section of ozone as function of temperature, in *Proceedings of the Quadrennial International Ozone Symposium*, J. London, ed., National Center for Atmospheric Research, Boulder, Colo., 140-145.

Bernanose, A. J., and M. G. Rene, 1959, Oxyluminescence of a few fluorescent compounds of ozone, *Adv. in Chem. Ser.* **21**:7-11.

Bhartia, P. K., K. F. Klenk, A. J. Fleig, C. G. Wellemeyer, and D. Gordon, 1984, Intercomparison of Nimbus 7 solar backscattered ultraviolet ozone profiles with rocket, balloon and Umkehr profiles, to be published in *J. Geophys. Res.* **89**:5527-5238.

Bojkov, R. D., 1969, Differences in Dobson spectrophotometer and filter ozonometer measurements of total ozone. *J. Appl. Met.* **8**:362-368.

Bol'shakova, L. G., 1976, Correction of atmospheric ozone content measurements for the effect of spectral-interval width, *Akad. Nauk SSSR Izv. Atmos. and Oceanic Phys.* **12**:969-978.

Bol'shakova, L. G., 1979, Determining the extraatmospheric intensity of solar radiation by the Bouguer method for ozonometers with narrow-band optical filters, *Akad. Nauk SSSR Izv. Atmos. and Oceanic Phys.* **15**:726-731.

Bowman, Lloyd D., and Richard F. Horak, 1974, *A Continuous Ultraviolet Absorption Ozone Photometer*, Air Quality Instrumentation, vol. 2, J. W. Scales, ed., Instr. Soc. of Am., 305-315.

Brewer, A. W., 1973, A replacement for the Dobson spectrophotometer, *Pure Appl. Geophys.* **106-108**:919-927.

Brewer, A. W., and J. B. Kerr, 1973, Total ozone measurements in cloudy weather, *Pure Appl. Geophys.* V-VII, **106-108**:928-937.

Brewer, A. W., and J. R. Milford, 1960, The Oxford-Kew ozonesonde, *R. Soc. (London) Proc.*, ser. A, **256**:460-495.

Brewer, A. W., H. U. Dütsch, J. R. Milford, M. Migeotte, H. K. Paetzold, F. Piscalar, and E. Vigroux, 1960, Distribution verticale de l'ozone atmosphérique comparaison de diverses méthodes, *Extraits Ann. de Geophys.* **16**:196-222.

Brezgin, N. I., G. I. Kuznetsov, A. F. Chizhov, and O. V. Shtyrkov, 1977, Rocket observations of atmospheric ozone and aerosol, in *Proceedings of the Joint Symposium on Atmospheric Ozone*, vol. 2, K. E. Grasnick, ed., Berlin, 47-48.

Browell, E. V., A. F. Carter, and S. T. Shipley, 1981, An airborne lidar system for ozone and aerosol profiling in the troposphere and lower stratosphere, in *Proceedings of the Quadrennial International Ozone Symposium*, J. London, ed., National Center for Atmospheric Research, Boulder, Colo., 99-107.

Cabannes, J., and J. Dufay, 1927, Les variations de la quantité d'ozone contenue dans l'atmosphère, *J. Phys. (Paris)* **8**:353-364.

Carver, J. H., B. H. Horton and F. G. Burger, 1966, Nocturnal ozone distribution in the upper atmosphere, *J. Geophys. Res.* **71**:4189-4191.

Chappuis, M. J., 1880, Sur le spectre d'absorption de l'ozone, *Acad. Sci. (Paris) C. R.* **91**:985-986.

Chappuis, M. J., 1882, Sur les spectres d'absorption de l'ozone et de l'acid pernitrique, *J. Phys. (Paris)*, 2nd ser; **1**:494-504; **2**:484.

Cornu, A., 1879, Observation de la limite ultraviolette du spectre solaire à diverses altitudes, *Acad. Sci. (Paris) C. R.* **89**:808-814.

Craig, R. A., 1965, *The Upper Atmosphere, Meteorology and Physics,* Academic Press, N.Y. and London.

Craig, R. A., 1976, Umkehr measurements at Tallahassee, *J. Appl. Meteorol.* **15**:509-513.

Craig, R. A., J. J. DeLuisi, and I. Stretrer, 1967, Comparison of chemiluminescent and Umkehr observations of ozone, *J. Geophys. Res.* **72**:1667-1671.

Crosby, D. S., W. G. Planet, A. J. Miller, and R. M. Nagatani, 1981, Evaluation and comparison of total ozone fields derived from TOVS and SBUV, in *Proceedings of the Quadrennial International Ozone Symposium,* J. London, ed., National Center for Atmospheric Research, Boulder, Colo., 161-167.

Crutzen, Paul J., and D. E. Ehhalt, 1977, Effects of nitrogen fertilizers and combustion on the stratospheric ozone layer, *Ambio* **6**:2-3, 112-117.

Cunnold, D. M., F. N. Alyea, and R. G. Prinn, 1980, Preliminary calculations concerning the maintenance of the zonal mean ozone distribution in the Northern Hemisphere, *Pure Appl. Geophys.* **118**:329-354.

Cunnold, D. M., F. N. Alyea, N. Phillips, and R. G. Prinn, 1975, A three-dimensional dynamical-chemical model of atmospheric ozone, *J. Atmos. Sci.* **32**:170-194.

Dave, J. V., and C. L. Mateer, 1967, A preliminary study on the possibility of estimating total atmospheric ozone from satellite measurements, *J. Atmos. Sci.* **24**:414-427.

Dave, J. V., P. A. Sheppard, and C. D. Walshaw, 1960, Ozone distribution and the continuum from observations in the region of the 1,043 cm^{-1} band, *R. Meteorol. Soc. Q. J.* **89**:307-318.

DeLuisi, J. J., 1975, Measurements of the extraterrestrial solar radiant flux from 2981-4000 Å and its transmission through the earth's atmosphere as it is effected by dust and ozone, *J. Geophys. Res.* **80**:345-354.

DeLuisi, J. J., 1978, Use of the Dobson spectrophotometer in a variable wavelength mode as an alternative to the Umkehr measurement for observing trends in the vertical distribution of ozone, in *Proceedings WMO Symposium on the Geophysical Aspects and Consequences of Changes in the Composition of the Stratosphere* (Toronto), WMO No. 511, Geneva, 125-130.

DeLuisi, J. J., 1979, Umkehr vertical ozone profile errors caused by the presence of stratospheric aerosols, *J. Geophys. Res.* **84**:1766-1770.

DeLuisi, J. J., and J. Nimira, 1978, Preliminary comparison of satellite BUV and surface-based Umkehr observations of the vertical distribution of ozone in the upper stratosphere, *J. Geophys. Res.* **83**:379-384.

DeLuisi, J. J., C. L. Mateer, and D. F. Heath, 1979, Comparison of seasonal variations of upper stratospheric ozone concentrations revealed by Umkehr and Nimbus 4 BUV observations, *J. Geophys. Res.* **84**:3728-3732.

Dierich, P., N. Monnanteuil, J. M. Colmont, A. Baudry, and J. de la Noe, 1981, A new millimeter facility in operation at the Bordeaux Observatory used for ground based survey of ozone, in *Proceedings of the Quadrennial International Ozone Symposium,* J. London, ed., National Center for Atmospheric Research, Boulder, Colo., 1971.

Dobson, G. M. B., 1930, Observations of the amount of ozone in the earth's atmosphere and its relation to other geophysical conditions, Part IV, *R. Soc. (London) Proc.*, ser. A, **129**:411-433.

Dobson, G. M. B., 1931, A photoelectric spectrophotometer for measuring the amount of atmospheric ozone, *Phys. Soc. (London) Proc.* **43**:324-339.

Dobson, G. M. B., 1957, Observers handbook for the ozone spectrophotometer, in *Ann. International Geophysical Year,* vol. 5, Pergamon, New York, 46-89.

Dobson, G. M. B., 1963, Notes on the measurement of ozone in the atmosphere, *R. Meteorol. Soc. Q. J.* **89**:409-411.

Dobson, G. M. B., 1973, Atmospheric ozone and the movement of the air in the stratosphere, *Pure Appl. Geophys.* **106-108**:1520-1530.

Dobson, G. M. B., and D. N. Harrison, 1926, Measurement of the amount of ozone in the earth's atmosphere and its relation to other geophysical conditions, Part I, *R. Soc. (London) Proc.*, ser. A, **110**:660-693.

Dobson, G. M. B., D. C. Harrison, and J. Lawrence, 1927, Measurement of the amount of ozone in the earth's atmosphere and its relation to other geophysical conditions, Part II, *R. Soc. (London) Proc.*, ser. A, **114**:521-541.

Dobson, G. M. B., D. C. Harrison, and J. Lawrence, 1929, Measurement of the amount of ozone in the earth's atmosphere and its relation to other geophysical conditions, Part III, *R. Soc. (London) Proc.*, ser. A, **122**:456-486.

Durand, E., 1949, Rocketsonde research at the Naval Research Laboratory, in *The Atmospheres of the Earth and Planets,* G. P. Kuiper, ed., Univ. of Chicago Press, Chicago, 134-148.

Dütsch, H. U., 1969, Atmospheric ozone and ultraviolet radiation, *World Survey of Climatology,* vol. 4, D. F. Rex, ed., Elsevier, New York, 383-432.

Dütsch, H. U., 1974, Ozone distribution in the atmosphere, *Can. J. Chem.* **52**:1491-1504.

Dütsch, H. U., 1978, Vertical ozone distribution on a global scale, *Pure Appl. Geophys.* **116**:511-529.

Dütsch, H. U., 1980, Vertical ozone distribution and tropospheric ozone, in *Proceedings of the NATO Advanced Study Institute on Atmospheric Ozone: Its Variation and Human Influences,* A. C. Aikin, ed., Rept. No. FAA-EE-80-20, U.S. Department of Transportation, Federal Aviation Administration, Washington, D.C., 7-30.

Dütsch, H. U., and Ch. Ch. Ling, 1969, Critical comparison of the determination of vertical ozone distribution by the Umkehr method and by electrochemical method, *Ann. Geophys.* **25**:211-214.

Dütsch, H. U., and Ch. Ch. Ling, 1973, Fourteen-year series of vertical ozone distribution over Arosa, Switzerland, from Umkehr measurements, *Pure Appl. Geophys.* **106-108**:1139-1150.

Dziewulska-Losiowa, A., and C. D. Walshaw, 1975, *The International Comparison of Ozone Spectrophotometers at Belsk,* 24 June-6 July, 1974, Institute of Geophysics, Polish Academy of Science, Warsaw.

Else, C. V., D. B. B. Powell, and E. L. Simmons, 1969, An improved solid state amplifier for Dobson ozone spectrophotometer, *Ann. Geophys.* **25**:313-315.

Evans, W. F. J., and E. J. Llewellyn, 1972, Molecular oxygen emission in the air glow, *Ann. Geophys.* **26**:167-178.

Evans, W. F. J., I. A. Asbridge, J. B. Kerr, C. L. Mateer, and R. A. Olafson, 1981,

The effects of SO_2 on Dobson and Brewer total ozone measurements, in *Proceedings of the Quadrennial International Ozone Symposium*, J. London, ed., National Center for Atmospheric Research, Boulder, Colo., 48-56.

Fabry, C., and M. Buisson, 1913, L'absorption de l'ultraviolet par l'ozone et la limite du spectre solaire, *J. Phys.* 3:196.

Fabry, C., and M. Buisson, 1921, Etude de l'extrémité ultraviolet du spectre solaire, *J. Phys. Rad.* 2:197-226.

Falconer, Phillip D., and James D. Holdeman, 1976, Measurements of atmospheric ozone made from a GASP-equipped 747 airliner: mid March, 1975, *Geophys. Res. Lett.* 3:101-104.

Feister, U., 1980, The determination of atmospheric total ozone using infrared radiation measurements with the Fourier spectrometer SI-1 onboard METEOR 28, *Z. Meteorologie* 5:279-295.

Feister, U., and D. Spänkuch, 1981, Total ozone retrieval from satellite METEOR 28 Fourier spectrometer measurements, in *Proceedings of the Quadrennial International Ozone Symposium*, J. London, ed., National Center for Atmospheric Research, Boulder, Colo., 168-175.

Fleig, Albert J., V. G. Kaveeshwar, K. F. Klenk, M. R. Hinman, P. K. Bhartia, and P. M. Smith, 1981, Characteristics of space and ground based total ozone observing systems investigated by intercomparison of Nimbus 4 backscattered ultraviolet (BUV) data with Dobson and M-83 results, in *Proceedings of the Quadrennial International Ozone Symposium*, J. London, ed., National Center for Atmospheric Research, Boulder, Colo., 9-16.

Fleig, Albert J., K. F. Klenk, P. K. Bhartia, D. Gordon, and W. S. Schneider, 1982, *User's Guide for the Solar Backscattered Ultraviolet (SBUV) Instrument, 1st-Year Ozone-S Data Set*, NASA Ref. Publ. 1095, Goddard Space Flight Center (NASA), Greenbelt, Maryland, 66p.

Frederick, J. E., R. B. Abrams, R. Dasgupta, and B. Guenther, 1980, An observed annual cycle in tropical upper stratospheric and mesospheric ozone, *Geophys. Res. Lett.* 7:713-716.

Frederick, J. E., R. B. Abrams, R. Dasgupta, and B. Guenther, 1981, Natural variability of tropical upper stratospheric ozone inferred from the atmospheric explorer backscatter ultraviolet experiment, *J. Atm. Sci.* 38:1092-1099.

Frederick, J. E., F. T. Huang, A. R. Douglass, and C. A. Reber, 1983, The distribution and annual cycle of ozone in the upper stratosphere, to be published in *J. Geophys. Res.* 6:3819-3828.

Funk, J. P., and G. L. Garnham, 1962, Australian ozone observations and a suggested 24-month cycle, *Tellus* 14:378-382.

Garrison, L. M., D. D. Doda, and A. E. S. Green, 1979, Total ozone determination by spectroradiometry in the middle ultraviolet, *Appl. Optics* 18:850-855.

Gerlach, J. C., and C. L. Parsons, 1981, Assessment of the problems in the use of interference filter photometers to determine total ozone, in *Proceedings of the Quadrennial International Ozone Symposium*, J. London, ed., National Center for Atmospheric Research, Boulder, Colo., 204-211.

Ghazi, Anver, 1980, *Atlas der Globalverteilung des Gesamtozonbetrages nach Satellitenmessungen* (April 1970-Mai 1972), Mitteilungen der Institut für Geophysik und Meteorologie der Universität zu Köln.

Gille, J. C., 1980, Ozone distributions by infrared limb scanning: preliminary results from the LRIR, in *Proceedings of the NATO Advanced Study Institute on Atmospheric Ozone: Its Variations and Human Influences,* A. C. Aikin, ed., Rep. no. FAA-EE-80-20, U.S. Department of Transportation, Federal Aviation Administration, Washington, D.C., 103-121.

Gille, John C., and F. B. House, 1971, On the inversion of limb radiance measurements, I: Temperature and thickness, *J. Atmos. Sci.* 28:1427-1442.

Gille, J. C., P. Bailey, F. B. House, R. A. Craig, and J. R. Thomas, 1975, The limb radiance inversion radiometer (LRIR) experiment, in *The NIMBUS 6 User's Guide,* Goddard Space Flight Center (NASA), Greenbelt, Maryland, 141-161.

Gille, J. C., G. P. Anderson, W. J. Kohri, and P. L. Bailey, 1981, Observations of the interaction of ozone and dynamics, in *Proceedings of the Quadrennial International Ozone Symposium,* J. London, ed., National Center for Atmospheric Research, Boulder, Colo., 1007-1011.

Goody, R. M., and W. T. Roach, 1956, Determination of the vertical distribution of ozone from emission spectra, *R. Meteorol. Soc. Q. J.* 82:217-221.

Götz, F. W. P., 1931, Zum Strahlungsklima des Spitzbergensommers, Strahlungs- und Ozonmesungen in der Konigsbucht, 1929, *Gerl. Beitr. Geoph.* 31:119-154.

Götz, F. W. P., A. R. Meetham, and G. M. B. Dobson, 1934, The vertical distribution of ozone in the atmosphere, *Roy. Soc. Proc.,* ser. A, 145:416-446.

Greenstone, R., 1978, The possibility that changes in cloudiness will compensate for changes in ozone and lead to natural protection against ultraviolet radiation, *J. Appl. Meteor.* 17:107-109.

Gushchin, G. P., ed., 1974, *Actionmetry, Atmospheric Optics, Ozonometry,* translated from the Russian, Keter Publishing House Jerusalem.

Gushchin, G. P., 1977, On the technique for measuring of atmospheric ozone at the world network of stations, in *Proceedings of the Joint Symposium on Atmospheric Ozone,* vol. 1, K. E. Grasnick, ed., Berlin, 135-147.

Hanser, Frederick, A., Bach Sellers, and Daniel C. Briehl, 1978, Ultraviolet spectrophotometer for measuring columnar atmospheric ozone from aircraft, *Appl. Optics* 17:1649-1656.

Harrison, I. D. N., and G. M. B. Dobson, 1925, Measurements of the amount of ozone in the upper atmosphere, *R. Meteorol. Soc. Q. J.* 51:363-369.

Hartley, W. N., 1880a, On the absorption spectrum of ozone, *Chem. News* 42:268.

Hartley, W. N., 1880b, On the probable absorption of solar radiation by atmospheric ozone, *Chem. News* 42:268.

Hartley, W. N., 1881a, On the absorption spectrum of ozone, *J. Chem. Soc.* 39:57-60.

Hartley, W. N., 1881b, The absorption of solar rays by atmospheric ozone, *J. Chem. Soc.* 39:111-128.

Hays, P. B., and R. G. Roble, 1973, Observation of mesospheric ozone at low latitudes, *Planet. Space Sci.* 21:273-279.

Heaps, W. S., T. J. McGee, R. D. Hudson, and L. O. Caudill, 1981, *Balloon Borne Lidar Measurements of Stratospheric Hydroxyl and Ozone,* Goddard Space Flight Center (NASA), Greenbelt, Maryland, X-963-81-27, 72p.

Heath, D. F., A. J. Krueger, and P. J. Crutzen, 1977, Solar proton event: influence on stratospheric ozone, *Science* 197:886-889.

Heath, D. F., Carlton L. Mateer, and Arlin J. Krueger, 1973, The Nimbus-4 backscatter ultraviolet (BUV) atmospheric ozone experiment—two years operation, *Pure Appl. Geophys.* **106-108**:1238-1253.

Heath, D. F., A. J. Krueger, H. A. Roeder, and B. D. Henderson, 1975, The solar backscatter ultraviolet and total ozone mapping spectrometer (SBUV/TOMS) for Nimbus-G, *Opt. Eng.* **14**:323-331.

Hering, Wayne S., 1966, Ozone and atmospheric transport processes, *Tellus* **18**(2):329-336.

Hering, W. S., and T. R. Borden, Jr., 1967, *Ozonesonde observations over North America*, vol. 4, AFCRL-64-30 (IV), Air Force Cambridge Research Laboratories, Bedford, Massachusetts, 365p.

Hering, W. S., and H. U. Dütsch, 1965, Comparison of chemiluminescent and electrochemical ozonesonde observations, *J. Geophys. Res.* **70**:5483-5490.

Hilsenrath, E., and P. T. Kirschner, 1980, Recent assessment of the performance and accuracy of a chemiluminescent rocketsonde for upper atmospheric ozone measurements, *Rev. Sci. Instr.* **51**:1381-1389.

Hilsenrath, E., and B. M. Schlesinger, 1981, Total ozone seasonal and interannual variations derived from the 7 year Nimbus-4 BUV data set, *J. Geophys. Res.* **86**:12087-12086.

Hilsenrath, Ernest, Lester Seiden, and Philip Goodman, 1969, An ozone measurement in the mesopshere and stratosphere by means of a rocketsonde, *J. Geophys. Res.* **74**:6873-6880.

Holton, J. R., and H. C. Tan, 1980, The influence of the equatorial quasi-biennial oscillation on the global circulation at 50 mb, *J. Atmos. Sci.* **37**:2200-2208.

Ilyas, M., 1981, Temporal variability and latitudinal asymmetry in the Woomera (31°S) ozone data, in *Proceedings of the Quadrennial International Ozone Symposium*, J. London, ed., National Center for Atmospheric Research, Boulder, Colo., 526-533.

Iozenas, V. A., V. A. Krasnopol'skiy, A. P. Kuznetsov, and A. I. Lebedinskiy, 1969, An investigation of the planetary ozone distribution from satellite measurements of ultraviolet spectra, *Akad. Nauk SSSR Izv. Atmos. Oceanic Phys.* **5**:219.

Johnson, F. S., J. D. Purcell, and R. Tousey, 1951, Measurements of the vertical distribution of atmospheric ozone from rockets, *J. Geophys. Res.* **56**:583-594.

Kay, R. H., 1954, The measurement of ozone vertical distribution by a chemical method to heights of 12 km from aircraft, in *Rocket Exploration of the Upper Atmosphere*, R. L. F. Boyd and M. J. Seaton, eds., Pergamon, London, 208-211.

Kay, R. H., A. W. Brewer, and G. M. B. Dobson, 1956, Some measurements of the vertical distribution of atmospheric ozone by a chemical method to heights of 15 km from aircraft, in *Scientific Proceedings of the International Association of Meteorology, Publ. A1M No. 10* (Rome, 1954), Butterworth Publications, Ltd., London, 189-193.

Keating, G. M., L. R. Lake, and J. V. Nicholson III, 1981, Global ozone—solar activity relationship from satellite measurements, in *Proceedings of the Quadrennial International Ozone Symposium*, J. London, ed., National Center for Atmospheric Research, Boulder, Colo., 1075-1082.

Kerr, J. B., C. T. McElroy, and R. A. Olafson, 1981, Measurements of ozone with the Brewer ozone spectrophotometer, in *Proceedings of the Quadrennial International Ozone Symposium*, J. London, ed., National Center for Atmospheric Research, Boulder, Colo., 74-79.

Khrgian, A. K., 1975, *The Physics of Atmospheric Ozone*, Israel Program for Scientific Translations, Jerusalem.

Klenk, Kenneth F., 1980, Absorption coefficients of ozone for the backscatter UV experiment, *Appl. Optics* 19:236-242.

Klenk, K. F., P. K. Bhartia, A. J. Fleig, V. G. Kaveeshwar, R. D. McPeters, and P. M. Smith, 1982, Total ozone determination from the backscattered ultraviolet (BUV) experiment, *J. App. Meteorol.* 21:1672-1684.

Knittel, Jürgen, 1976, *Ein Beitrag zur Klimatologie der Stratosphäre der Südhalbkugel*, Meteorol. Abh. ser. A. mono., Dietrich Reimer, Berlin.

Kobayashi, J., and Y. Toyama, 1966, On various methods of measuring the vertical distribution of atmospheric ozone (II), *Papers in Meteorol. and Geophys.* 17:97-112.

Komhyr, W. D., 1969, Electrochemical concentration cells for gas analysis, *Ann. Geophys.* 25:203-210.

Komhyr, W. D., 1980a, *Operations Handbook—Ozone Observations with a Dobson Spectrophotometer*, WMO Global Ozone Research and Monitoring Project, Rept. 6, Geneva.

Komhyr, W. D., 1980b, Dobson spectrophotometer systematic total ozone measurement error., *Geophys. Res. Lett.* 7:161-163, February 1980.

Komhyr, W. D., and R. D. Evans, 1980, Dobson spectrophotometer total ozone measurement errors caused by interfering absorbing species such as SO_2, NO_2, and photochemically produced O_3 in polluted air, *Geophys. Res. Lett.* 7:157-160.

Komhyr, W. D., and R. D. Grass, 1972, Dobson ozone spectrophotometer modification, *J. Appl. Meteorol.* 11:858-863.

Komhyr, W. D., and T. B. Harris, 1971, *Development of an ECC Ozonesonde*, Tech. Rep., ERL 200-ADCL 18, National Oceanic and Atmospheric Administration, Boulder, Colo.

Komhyr, W. D., R. D. Grass, and R. K. Leonard, 1981, International comparison of Dobson ozone spectrophotometers, in *Proceedings of the Quadrennial International Ozone Symposium*, J. London, ed., National Center for Atmospheric Research, Boulder, Colo., 25-32.

Komhyr, W. D., E. W. Barrett, G. Slocum, and H. K. Weickmann, 1971, Atmospheric total ozone increase during the 1960s, *Nature* 232(5310):390-391.

Konkov, V. I., V. A. Kononkov, and S. P. Perov, 1981, A chemiluminescent method of rocket measurement of ozone vertical distribution, *Symposium on Middle Atmosphere Science*, vol. 22, International Association of Meteorology and Atmospheric Physics, Hamburg.

Krueger, A. J., 1965, Rocket ozonesonde (Rocoz), in *Proceedings of the Atmospheric Ozone Symposium*, H. U. Dütsch, ed., Albuquerque, N. Mex., WMO, Geneva, 24-26.

Krueger, A. J., 1973, The mean ozone distribution from several series of rocket soundings to 52 km at latitudes from 58°S to 64°N, *Pure Appl. Geophys.* 106-108:1272-1280.

Krueger, A. J., and W. R. McBride, 1968, *Sounding Rocket OGO-4 Satellite Ozone Experiment: Rocket Ozonesonde Measurements,* TP 4667, Naval Weapons Center, China Lake, Calif.

Kulke, W., and H. K. Paetzold, 1957, Über eine radiosonde zur bestimmung der vertikalen ozonverteilung, *Ann. Meteorol.* 8:47-53.

Kulkarni, R. N., and G. L. Garnham, 1970, Longitudinal variation of ozone in the lower middle latitudes of the Southern Hemisphere, *Commonwealth Sci. Industrial Res. Org.* 75:4144-4176.

Kulkarni, R. N., and A. B. Pittock, 1970, Results of a comparison between Umkehr and ozonesonde data, *R. Meteorol. Soc. Q. J.* 96:739-743.

Kunzi, K. F., and J. Fulde, 1977, Passive microwave probing of the earth atmosphere: A new tool for ozone research, in *Proceedings of the Joint Symposium on Atmospheric Ozone,* vol. 1, K. E. Grasnick, ed., Berlin, 203-206.

Kuznetsov, G. I., 1977, New multiwave method and instrument for observation of atmospheric ozone and aerosol, *Pol. Acad. Sci. Inst. Geophys.* 90:13-20.

Kuznetsov, G. I., A. G. Chizhov, and O. V. Shtyrkov, 1975, The optical investigation of vertical distribution of scattering coefficient and ozone in upper atmosphere by meteorological rocket MP-12, presented at *XVIII COSPAR Plenary Meeting Varna* (Bulgaria).

Ladenburg, R., 1929, Bemerkungen über den Absorptionkoeffizienten des ozons, *Ger. Beträge zur Geophys.* 24:40-41.

Lean, J. L., Observation of the diurnal variation of atmospheric ozone, *J. Geophys. Res.* 87:4973-4980.

List, R. J., 1968, *Smithsonian Meteorological Tables,* 6th ed., Smithsonian Institution Press, Washington, D.C.

London, J., 1963, The distribution of total ozone in the Northern Hemisphere, *Beitr. Phys. Atmos.* 36:254-263.

London, J., 1967, The average distribution and time variation of ozone in the stratosphere and mesosphere, *Space Res. VII,* R. L. Smith-Rose, ed., North Holland, Amsterdam, 172-185.

London, J., 1978a, Distribution of atmospheric ozone and how it is measured, in *Air Quality Meteorology and Atmospheric Ozone,* A. L. Morris and R. C. Barras, eds., ASTM STP 653, American Society for Testing and Materials, Philadelphia, 339-364.

London, J., 1978b, Long period time variations of ozone in the lower stratosphere, in *Proceedings of the 4th Joint Conference on Sensing of Environmental Pollutants,* American Chemical Society, Washington, D.C., 677-680.

London, J., 1980, The observed distribution and variations of total ozone, in *Proceedings of the NATO Advanced Study Institute on Atmospheric Ozone: Its Variation and Human Influences,* A. C. Aikin, ed., Rep. No. FA-EE-80-20, U.S. Department of Transportation, Federal Aviation Administration, Washington, D.C., 31-44.

London, J., 1983, Periodic and aperiodic ozone variations in the middle and upper stratosphere, *Adv. Space Res.* 2:201-204.

London, J., and J. K. Angell, 1982, The observed distribution of ozone and its variations, in *Stratospheric Ozone and Man,* F. A. Bower and R. B. Ward, eds., Chemical Rubber Co., CRC Press, Inc., Boca Raton, Florida, 7-42.

London, J., and J. Kelley, 1974, Global trends in atmospheric ozone, *Science* **184**:987-989.

London, J., and Xiude Ling, 1981, The geographic bias in determining average variations of total ozone from ground-based observations, in *Proceedings of the Quadrennial International Ozone Symposium*, J. London, ed., National Center for Atmospheric Research, Boulder, Colo., 337-339.

London, J., J. E. Frederick, and G. P. Anderson, 1977, Satellite observations of the global distribution of stratospheric ozone, *J. Geophys. Res.* **82**:2543-2556.

London, J., J. E. Frederick, and G. P. Anderson, 1977, Satellite observation of stratospheric ozone, in *Proceedings of the Joint IAOC/ICACGP Symposium on Atmospheric Ozone*, International Association of Meteorology and Atmospheric Physics/International Union of Geodesy and Geophysics, August 1976, Vol. 1, Dresden, GDR, 387-403.

London, J., R. D. Bojkov, S. Oltmans, and J. I. Kelley, 1967, *Atlas of the Global Distribution of Total Ozone, July 1957-June 1967*, National Center for Atmospheric Research, Boulder, Colo.

Lovill, J. E., T. J. Sullivan, R. L. Weichel, J. S. Ellis, J. G. Huebel, J. A. Korver, P. P. Weidhass, and F. A. Phelps, 1978, *Total Ozone Retrieval from Satellite Multichannel Filter Radiometer Measurements*, UCRL-52473, University of California/Livermore, Lawrence Livermore Laboratory 97p.

Luther, F. M., and R. L. Weichel, 1981, Determination of total ozone from DMSP multichannel filter radiometer measurements, in *Proceedings of the Quadrennial International Ozone Symposium*, J. London, ed., National Center for Atmospheric Research, Boulder, Colo., 17-24.

McCormick, M. P., Patrick Hamill, T. J. Pepin, W. P. Chu, T. S. Swissler, and L. R. McMaster, 1979, Satellite studies of the stratospheric aerosol, *Am. Meteorol. Soc. Bull.* **60**:1038-1046.

Mahlman, J. D., D. G. Andrews, H. U. Dütsch, D. L. Hartmann, T. Matsuno, R. J. Murgatroyd, and J. F. Noxon, 1981, Transport of trace constituents in the stratosphere, in *Middle Atmosphere Program, Handbook for MAP*, vol. 3, C. F. Sechrist, Jr., ed., SCOSTEP Secretariat, University of Illinois, Urbana, 14-43.

Mateer, C. L., 1965, On the information content of Umkehr observations, *J. Atmos. Sci.* **22**:370-381.

Mateer, C. L., 1981, A review of some unresolved problems in the measurement/estimation of total ozone and the vertical ozone profile, in *Proceedings of the Quadrennial International Ozone Symposium*, J. London, ed., National Center for Atmospheric Research, Boulder, Colo., 1-8.

Mateer, C. L., and I. A. Ashbrige, 1981, On the appropriate haze correction for direct sun total ozone measurements with the Dobson spectrophotometer, in *Proceedings of the Quadrennial International Ozone Symposium*, J. London, ed., National Center for Atmospheric Research, Boulder, Colo., 236-242.

Mateer, C. L., and J. J. DeLuisi, 1981, The estimation of the vertical distribution of ozone by the short Umkehr method, in *Proceedings of the Quadrennial International Ozone Symposium*, J. London, ed., National Center for Atmospheric Research, Boulder, Colo., 64-73.

Mateer, C. L., and H. U. Dütsch, 1964, *Uniform Evaluation of Umkehr Observations, Part I*, National Center for Atmospheric Research, Boulder, Colo.

Mateer, C. L., D. F. Heath, and A. J. Krueger, 1971, Estimation of total ozone from satellite measurements of backscattered ultraviolet earth radiance, *J. Atmos. Sci.* **28**:1307-1311.

Matthews, W. A., and P. Fabian, 1981, Scanning filter photometer for O_3-NO_2 column abundance measurements, in *Proceedings of the Quadrennial International Ozone Symposium*, J. London, ed., National Center for Atmospheric Research, Boulder, Colo., 137-139.

Matthews, W. A., R. E. Basher, and G. J. Fraser, 1974, Filter ozone spectrophotometer, *Pure Appl. Geophys.* **112**:931-938.

Menzies, R. T., and R. K. Seals, Jr., 1977, Ozone monitoring with an infrared heterodyne radiometer, *Science* **197**:1275-1277.

Miller, A. J., B. Korty, E. Hilsenrath, A. J. Fleig, and D. F. Heath, 1978, Verification of Nimbus 4 BUV total ozone data and the requirements for operational satellite monitoring, in *Proceedings of the WMO Symposium on the Geophysical Aspects and Consequences of Changes in the Composition of the Stratosphere*, Toronto, WMO No. 511, Secretariat of the World Meteorological Organization, Geneva, 153-160.

Miller, A. J., R. M. Nagatani, J. D. Laver, and B. Korty, 1979, Utilization 100 mb mid-latitude height fields as an indicator of sampling effects on total ozone variations, *Mon. Weather Rev.* **107**:782-787.

Miller, A. J., T. G. Rogers, R. M. Nagatani, D. F. Heath, A. J. Fleig, and V. G. Kaveeshwar, 1981, Results and analyses of ground-based and satellite ozone observations—a review, in *Proceedings of the Quadrennial International Ozone Symposium*, J. London, ed., National Center for Atmospheric Research, Boulder, Colo., 285-297.

Millier, F., B. A. Emery, and R. G. Roble, 1981, OSO-8 lower mesospheric ozone number density profiles, in *Proceedings of the Quadrennial International Ozone Symposium*, J. London, ed., National Center for Atmospheric Research, Boulder, Colo., 572-575.

Millier, F., A. Vidal-Madjar, J. Guidon, and R. G. Roble, 1979, Ozone number density profiles in the lower mesosphere as determined by the French experiment onboard OSO-8, *Geophys. Res. Lett.* **6**:863-865.

Moxim, W. J., and J. D. Mahlman, 1978, Evaluation of various total ozone sampling networks using the GFDL 3-D tracer model, in *WMO Symposium on the Geophysical Aspects and Consequences of Changes in the Composition of the Stratosphere*, Toronto, Secretariat of the World Meteorological Organization, Geneva, No. 511, 217.

Moxim, W. J., and J. D. Mahlman, 1980, Evaluation of various total ozone sampling networks using the GFDL 3-D tracer model, *J. Geophys. Res.* **85**:4527-4539.

Murgatroyd, R. J., 1980, An introduction to studies of the general characteristics of the stratosphere and mesosphere, in *Proceedings of the NATO Advanced Study Institute on Atmospheric Ozone: Its Variations and Human Influences*, A. C. Aikin, ed., Rep. No. FAA-EE-80-20, U.S. Department of Transportation, Federal Aviation Administration, Washington, D.C., 689-701.

Newell, R. E., 1961, The transport of trace substances in the atmosphere and their implications for the general circulation of the stratosphere, *Geofis. Pura e Appl.* **49**:137-158.

Newell, R. E., 1964, Further ozone transport calculations and the spring maximum in ozone amount, *Pure Appl. Geophys.* **59**:191-206.

Nicolet, M., 1975, Stratospheric ozone: An introduction to its study, *Rev. Geophys. Space Phys.* **13**:593-636.

Ogawa, T., and T. Watanabe, 1981, Summary of the mesospheric ozone measurements during 1970-1979 in Japan, in *Proceedings of the Quadrennial International Ozone Symposium*, J. London, ed., National Center for Atmospheric Research, Boulder, Colo., 520-525.

Olafson, R. A., 1969, Mercury rectifier for the Dobson spectrophotometer, in *Symposium sur L'Ozone Atmosphérique*, A. Vassy, ed., 2-7 Sept. 1968, Centre National de la Recherche Scientifique, Monaco, 63.

Oltmans, S. J., and J. London, 1982, The quasi-biennial oscillation in atmospheric ozone, *J. Geophys. Res.* **87**:8981-8989.

Osherovich, A. L., M. Ya. Rozinskiy, and L. N. Yurganov, 1969, A study of atmospheric ozone during a total solar eclipse on September 22, 1968, *Akad. Nauk SSSR Izv. Atmos. and Oceanic Physics* **5**:1223-1226.

Paetzold, H. K., 1950, Eine Bestimmung der vertikalen Verteilung des atmosphärischen Ozons mit Hilfe von Monfinstenissen, *Z. Naturforschung* **5a**:661-666.

Paetzold, H. K., 1955, New experimental and theoretical investigations on the atmospheric ozone layer, *J. Atmos. Terr. Phys.* **7**:128-140.

Paetzold, H. K., 1973, The influence of solar activity on the stratospheric ozone layer, *Pure Appl. Geophys.* **106-108**(V-VII):1308-1311.

Paneth, F. A., and E. Gluckhauf, 1941, Measurement of atmospheric ozone by a quick electrochemical method, *Nature* **147**:614-615.

Parrish, A., R. deZafra, and P. Solomon, 1981, Ground-based mm-wave emission spectroscopy for the detection and monitoring of stratospheric ozone, in *Proceedings of the Quadrennial International Ozone Symposium*, J. London, ed., National Center for Atmospheric Research, Boulder, Colo., 122-129.

Parsons, C. L., J. C. Gerlach, and M. E. Williams, 1981, Preliminary results of an intercomparison of total ozone spectrophotometers, *Proceedings of the Quadrennial International Ozone Symposium*, J. London, ed., National Center for Atmospheric Research, Boulder, Colo., 80-87.

Pelon, J., P. Flamant, J. L. Chanin, and G. Megie, 1981, Vertical ozone profiles (5-25 km) as measured using the differential absorption lidar technique, *Proceedings of the Quadrennial International Ozone Symposium*, J. London, ed., National Center for Atmospheric Research, Boulder, Colo., 130-135.

Penndorf, R., 1948, Effects of the ozone shadow, *J. Meteorol.* **5**:152-160.

Pokrovskiy, O. M., and Ye. Ye. Kaygorodtsev, 1978, Information content of remote atmospheric ozone measurements, *Akad. Nauk. SSSR Izv. Atmos. and Oceanic Phys.* **14**:610-616.

Powell, D. B. B., 1971, The absorption coefficients of ozone for the Dobson spectrophotometer: A direct determination of their ratios and temperature dependence, *R. Meteorol. Soc. Q. J.* **97**:83-86.

Prabhakara, C., 1969, Feasibility of determining atmospheric ozone from outgoing infrared energy, *Mon. Weather Rev.* **97**:307-314.

Prabhakara, C., E. B. Rodgers, and V. V. Solomonson, 1973, Remote sensing of the global distribution of the total ozone and the inferred upper-tropospheric circulation from Nimbus IRIS, *Pure Appl. Geophys.* **106-108**:1226-1237.

Prabhakara, C., B. J. Conrath, R. A. Hanel, and E. J. Williamson, 1970, Remote sensing of atmospheric ozone using the 9.6 μm band, *J. Atmos. Sci.* 27:689-697.

Prabhakara, C., E. B. Rodgers, B. J. Conrath, R. A. Hanel, and V. G. Kunde, 1976, The Nimbus 4 infrared spectroscopy experiment 3. Observations of the lower stratospheric thermal structure and total ozone, *J. Geophys. Res.* 81:6391-6399.

Prior, E. J., and Bharat J. Oza, 1978, First comparison of simultaneous IRIS, BUV and ground-based measurements of total ozone, *Geophys. Res. Lett.* 5:547-550.

Raeber, Jost A., 1973, An automated Dobson spectrophotometer, *Pure Appl. Geophys.*, **106-108**:947-949.

Ramanathan, K. R., 1963, Biennial variation of atmospheric ozone over the tropics, *R. Meteorol. Soc. Q. J.* 89:540-542.

Ramanathan, K. R., P. D. Angreji, and G. M. Shah, 1969, The second Umkehr observed in zenith sky twilight and its interpretation, *Ann. Geophys.* 25:243-248.

Rangarajan, S., and A. Mani, 1981, Longitudinal anomalies in total ozone, in *Proceedings of the Quadrennial International Ozone Symposium*, J. London, ed., National Center for Atmospheric Research, Boulder, Colo., 418-420.

Rawcliffe, R. D., G. E. Meloy, R. M. Friedman, and E. H. Rogers, 1963, Measurements of vertical distribution of ozone from a polar orbiting satellite, *J. Geophys. Res.* 68:6425-6429.

Reed, Edith I., 1968, A night measurement of mesospheric ozone by observations of ultraviolet air glow, *J. Geophys. Res.* 73:2951-2957.

Reed, R. J., 1964, A tentative model of the 26-month oscillation in tropical latitudes, *R. Meteorol. Soc. Q. J.* 90:441-466.

Reed, R. J., W. J. Campbell, L. A. Rasmussen, and D. G. Rogers, 1961, Evidence of a downward-propagating, annual wind reversal in the equatorial stratosphere, *J. Geophys. Res.* 66:813-818.

Regener, V. H., 1954, *On the Vertical Distribution of Ozone,* Sci. Rept. 2, Contract AF 19(122)-381, University of New Mexico, Albuquerque, N. Mex., 54p.

Regener, V. H., 1960, On a sensitive method of the recording of atmospheric ozone, *J. Geophys. Res.* 65:3975-3977.

Regener, V. H., 1964, Measurement of atmospheric ozone with the chemiluminescent method, *J. Geophys. Res.* 69:3795-3800.

Regener, E., and V. H. Regener, 1934, Aufnahmen des ultravioletten Sonnenspecktrums in der Stratosphäre und die vertikale Ozonverteilung, *Physik. Z.* 35:788-793.

Reinsel, G., G. C. Tiao, M. N. Wang, R. Lewis, and D. Nychka, 1981, Statistical analysis of stratospheric ozone data for the detection of trends, *Atmos. Eng.* 15:1569-1577.

Riegler, G. R., J. F. Drake, S. C. Liu, and R. J. Cicerone, 1976, Stellar occultation measurements of atmospheric ozone and chlorine from OAO3, *J. Geophys. Res.* 81:497-501.

Riegler, G. R., S. K. Atreya, R. C. Cicerone, J. F. Drake, and S. C. Liu, 1977, Stellar occultation measurements of nighttime equatorial ozone between 42 km and 114 km altitude, in *Proceedings of the Joint Symposium on Atmospheric Ozone, vol. 2,* C. E. Grasnick, ed., Berlin, 7-19.

Robbins, D. E., and J. G. Carnes, 1978, Variations in the upper atmosphere's ozone profile, in *WMO Symposium on the Geophysical Aspects and Consequences of Changes in the Composition of the Stratosphere,* WMO no. 511, World Meteorological Organization, Geneva, 131-136.

Rusch, D. W., G. H. Mount, C. A. Barth, G. J. Rottman, R. J. Thomas, G. E. Thomas, R. W. Sanders, G. M. Lawrence, and R. S. Eckman, 1983, Ozone densities in the lower mesosphere measured by a limb scanning ultraviolet spectrometer, *Geophys. Res. Lett.* **10**:241-244.

Schönbein, C. F., 1840, Recherches sur la nature de l'odeur qui se manifeste dans certaines actions chimiques, *Acad. Sci. C. R.* **10**:706-710.

Secroun, C., A. Barbe, P. Marche, and P. Jouve, 1981, Application des techniques infrarouges à l'étude de l'ozone atmosphérique, in *Proceedings of the Quadrennial International Symposium,* J. London, ed., National Center for Atmospheric Research, Boulder, Colo., 108-111.

Sekihara, K., and C. D. Walshaw, 1969, The possibility of ozone measurements from satellites using the 1043 cm^{-1} band, *Ann. Geophys.* **25**:233-241.

Shimabukuro, F. I., P. L. Smith, and W. J. Wilson, 1975, Estimation of the ozone distribution from millimeter wavelength absorption measurements, *J. Geophys. Res.* **80**:2957-2959.

Shimabukuro, F. I., P. L. Smith, and W. J. Wilson, 1977, Estimation of the daytime and nighttime distribution of atmospheric ozone from ground-based millimeter wavelength measurements, *J. App. Meteorol.* **16**:929-934.

Shimizu, M., 1971, Global distribution and seasonal changes of total ozone amount in the atmosphere, *Geophys. Mag.* **35**:401-429.

Singer, S. F., 1956, Geophysical research with artificial earth satellites, *Adv. Geophys.* **3**:302.

Smith, W. L., H. M. Woolf, C. M. Hayden, D. Q. Wark, and L. M. McMillian, 1979, The TIROS-N operational vertical sounder, *Am. Meteorol. Soc. Bull.* **60**:1177-1187.

Sonntag, D., 1976, The electrochemical ozone sensor of the GDR for radiosondes and the evaluation of the measuring data, Abstract of Papers, in *Proceedings of the Joint Symposium on Atmospheric Ozone,* vol. 1, K. E. Grasnick, ed., Dresden, 175-178.

Sreedharan, C. R., 1968, An Indian electrochemical ozonesonde, *J. Sci. Instru.,* series 2, **1**:995-997.

Stair, R., and I. F. Hand, 1939, Methods and results of ozone measurements over Mount Evans, Colo., *Mon. Weather Rev.* **67**:331-338.

Strutt, R. J., 1918, Ultraviolet transparency of the lower atmosphere and its relative poverty in ozone, *R. Soc. (London) Proc.,* ser. A, **94**:260-268.

Subbaraya, B. H., and S. Lal, 1978, Rocket Measurements of Ozone Concentrations in the Equatorial Stratosphere at Thumba, paper presented at COSPAR Symposium, Innsbruck.

Sundararaman, N., T. Perry, Jr., W. Gurkin, E. Jackson, B. Horton, J. Lean, E. Llewellyn, B. Solheim, W. F. J. Evans, B. H. Subbaraya, S. Lal, T. Ogawa, T. Watanabe, E. Hilsenrath, and A. Krueger, 1981, International ozone rocket-sonde intercomparison, in *Proceedings of the Quadrennial International Ozone Symposium*, J. London, ed., National Center for Atmospheric Research, Boulder, Colo., 421-422.

Szwarc, V. S., 1981, An evaluation of the U.S.S.R. M-83 ozonometer, in *Proceedings of the Quadrennial International Ozone Symposium*, J. London, ed., National Center for Atmospheric Research, Boulder, Colo., 57-63.

Thomas, G. E., C. A. Barth, E. R. Hansen, C. W. Hord, G. M. Lawrence, G. H. Mount, G. J. Rottman, D. W. Rusch, A. I. Stewart, R. J. Thomas, J. London, P. L. Bailey, P. J. Crutzen, R. E. Dickinson, J. C. Gille, S. C. Liu, J. F. Noxon, and C. B. Farmer, 1980, Scientific objectives of the Solar Mesosphere Explorer Mission, *Pure Appl. Geophys.* 118:592-693.

Thomas, R. J., C. A. Barth, G. J. Rottman, D. W. Rusch, G. H. Mount, G. M. Lawrence, R. W. Sanders, G. E. Thomas, and L. E. Clemens, 1983a, Ozone density distribution in the mesosphere (50-90 km) measured by the SME limb scanning near infrared spectrometer, *Geophys. Res. Lett.* 10:245-248.

Thomas, R. J., C. A. Barth, G. J. Rottman, D. W. Rusch, G. H. Mount, G. M. Lawrence, R. W. Sanders, G. E. Thomas and L. E. Clemens, 1983b, Mesospheric ozone depletion during the solar proton event of July 13, 1982, Part I: Measurement, *Geophys. Res. Lett.* 10:253-255.

Timofeyev, Yu. M., and M. S. Biryulina, 1981, Combined use of measurements of outgoing UV and IR radiation for retrieving vertical profile and total content of ozone, *Akad. Nauk SSSR Izv. Atm. and Oceanic Phys.* 17:198-202.

Tohmatsu, T., 1977, Altitude distributions of minor atmospheric species in the mesosphere and lower thermosphere as measured in optical absorption and emission, *Space Res.* 17:247-251.

Tohmatsu, T., R. Ogawa, and T. Watanabe, 1974, Absorption by the upper atmosphere in the middle ultraviolet region, *Space Res.* 14:177-180.

Tolson, R. H., 1981, Quasi-biennial variations in zonal mean total columnar ozone derived from 7 years of BUV data, in *Proceedings of the Quadrennial International Ozone Symposium*, J. London, ed., National Center for Atmospheric Research, Boulder, Colo., 314-321.

Tucker, G. B., 1979, The observed zonal wind cycle in the southern hemisphere stratosphere, *R. Meteorol. Soc. Q. J.* 105:263-273.

Twomey, S., 1965, The application of numerical filtering to the solution of integral equations encountered in indirect sensing measurements, *Franklin Inst. J.* 279:95.

Van Loon, H., and R. L. Jenne, 1974, Standard deviations of monthly mean 500- and 100-mb heights in the Southern Hemisphere, *J. Geophys. Res.* 79:5661-5664.

Van Loon, H., K. Labitzke and R. L. Jenne, 1974, Standard deviations of 24-hour 10 mb height and temperature changes in the Northern Hemisphere, *Mon. Weather Rev.* 102:394-405.

Vassy, A., and S. I. Rasool, 1959, Un appareil simple pour la mesure rapide de l'epaisseur reduite de l'ozone atmosphérique, *J. Mecan. Phys. Atmos.* 1:109-117.

Venkateswaran, S. V., J. G. Moore, and A. J. Krueger, 1961, Determination of the vertical distribution of ozone by satellite photometry, *J. Geophys. Res.* **66**:1751-1771.

Vernazza, J., and P. Foukal, 1980, On the secular behavior of seasonal changes in ozone column density, *Geophys. Res. Lett.* **7**:993-994.

Vigroux, E., 1953, Contributions à l'étude experimentale de l'absorption de l'ozone, *Ann. Geophys.* **9**:709-762.

Vigroux, E., 1959, Distribution verticale de l'ozone atmosphérique d'àpres les observations de la bande, 9.6 μ, *Ann. Geophys.* **15**:516-538.

Vigroux, E., 1967, Détermination des coefficients moyens d'absorption de l'ozone en vue des observations concenant l'ozone atmosphérique à l'aide du spectromètre Dobson, *Ann. Phys. (Paris)* **2**:209-215.

Vigroux, E., 1971, Observations de routine de l'ozone. Comparaison de l'Umkehr à la méthode infra-rouge, *Ann. Geophys.* **27**(4):507-511.

Walshaw, C. D., 1960, The accuracy of determination of the vertical distribution of atmospheric ozone from emission spectrophotometry in the 1,043 cm^{-1} band at high resolution, *R. Meteorol. Soc. Q. J.* **86**:370.

Walshaw, C. D., G. M. B. Dobson, and B. M. F. MacGarry, 1971, Absorption coefficients of ozone for use in the determination of atmospheric ozone, *R. Meteorol. Soc. Q. J.* **97**:75-82.

Wardle, D. I., 1963, A new instrument for atmospheric ozone, *Nature* **199**(4899):1177-1178.

Waters, J. W., J. C. Hardy, R. F. Jarnot, and H. M. Pickett, 1981, Chlorine monoxide radical, ozone, and hydrogen peroxide: stratospheric measurements by microwave limb sounding, *Science* **214**:61-64.

Westbury, P. R., A. J. Thomas, and E. L. Simmons, 1981, An ergonomic design of electronics for the Dobson spectrophotometer, in *Proceedings of the Quadrennial International Ozone Symposium*, J. London, ed., National Center for Atmospheric Research, Boulder, Colo., 271-275.

Wilcox, R. W., G. D. Nastrom, and A. D. Belmont, 1977, Periodic variations of total ozone and its vertical distribution, *J. Appl. Meteorol.* **16**:290-298.

WMO, 1981, *Assessment of Performance Characteristics of Various Ozone Observing Systems, Report of the Meeting of Experts, Boulder, July 1980*, WMO Global Ozone Research and Monitoring Project, Report No. 9, 67p.

WMO, 1982, *The Stratosphere 1981, Theory and Measurements*, WMO Global Ozone Research and Monitoring Project, Rep. No. 11, Geneva.

Zerefos, C. S., 1983, On the quasi-biennial oscillation in equatorial stratospheric temperatures and total ozone, *Adv. Space Res.* **2**:177-181.

Zerefos, C. S., C. C. Repapis, and R. L. Jenne, 1982, Representativeness of total ozone trends as derived from satellite BUV and ground-based measurements, *Pure Appl. Geophys.* **120**:29-53.

Chapter 2

Ozone Photochemistry in the Stratosphere

Robert C. Whitten
NASA-Ames Research Center

Sheo S. Prasad
Jet Propulsion Laboratory
California Institute of Technology
and
Department of Physics
University of Southern California

FUNDAMENTAL MOLECULAR PROPERTIES

Ozone, a triatomic molecule that belongs to the C_{2v} point group, has a symmetrical bent structure in the ground state with an equilibrium apex angle $\theta = 116.8°$ and equilibrium distance $R_e = 1.271$ Å between the outer oxygen atom and the oxygen atom at the apex (Trambarulo et al., 1953; Kaplan, Migeotte, and Nevin, 1956; Hughes, 1956; Danti and Lord, 1959). Electron diffraction studies of Shand and Spurr (1943) also suggest a symmetrical bent structure. The heat of formation ΔH_f is 34.2 $\pm$ 0.4 kcal/mol at 298 K and 34.8 $\pm$ 0.4 kcal/mol at 0 K. Thus, this species is bound by about 25.3 kcal/mol or 1.1 eV. In chemical reactions, therefore, ozone readily gives up an oxygen atom, much more readily than other triatomic species in its class such as SO_2. Historically, there were suggestions of structures such as

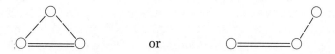

where an oxygen atom is attached at a relatively large distance to an almost intact oxygen molecule (see Wright, 1973 for additional discussions).

But the possibility of these structures for the ground state was eliminated by electron diffraction and microwave studies (Trambarulo et al., 1953; Kaplan, Migeotte, and Nevin, 1956; Hughes, 1956; Shand and Spurr, 1943).

The ground state of ozone can therefore be represented to a good approximation as three covalently bonded oxygen atoms. Since each oxygen atom has two unpaired electrons, the central oxygen atom can form a covalent bond with each of the two atoms, leaving an unpaired electron on each terminal atom. These unpaired electrons can be coupled into a singlet or triplet state and may occupy either pi (π) orbitals perpendicular to the molecular plane or sigma (σ) orbitals in the plane. The ground $^1A_1$ singlet state therefore contains two $p\pi$ electrons (π_a and π_b) coupled into a singlet. Thus, the ground state of ozone is well approximated as a bi-radical (Hayes and Siu, 1971; Hay, Dunning, and Goddard, 1975)

(paired nonbonding orbitals are not shown) with weak bonding between the singly occupied π orbitals on the terminal oxygen atom. In the valence bond notation this can be represented as

$$(l_a^2)\,(l_b^2)\,(\pi_c^2)\,(\sigma_b\overline{\sigma}_{c_1} + \sigma_{c_1}\overline{\sigma}_a)\,(\sigma_a\overline{\sigma}_{c_2} + \sigma_{c_2}\overline{\sigma}_b)\quad(\pi_a\pi_b + \pi_b\pi_a)$$

where l_a and l_b denote the lone pair $p\pi$ orbitals on the terminal atom, π_c is the lone pair ($p\pi$) orbital on the central atom, and $(\sigma_a\sigma_{c1})$ and $(\sigma_b\sigma_{c2})$ are sigma bonding orbitals on the two O—O bonds (Hay and Dunning, 1977). The lowest excited electronic stage of O_3 arises from the various ways of distributing atoms in the l_a, l_b, π_a, and π_b orbitals of the terminal atoms.

Several ab initio quantum-mechanical calculations have been devoted to the understanding of the electronic structure of ozone (Hay, Dunning, and Goddard, 1975; Hay et al., 1982; Wilson and Hopper, 1981; Thuneman, Peyerimhof, and Buenker, 1978; Lucchese and Schaeffer, 1977; Shih, Buenker, and Peyerimhof, 1974, and references to earlier studies cited therein). The ground state potenital energy surface has also been studied in some detail (Wilson and Hopper, 1981; Wright, Shih, and Buenker, 1980; Murrell, Sorbie, and Marandas, 1976; and Carney, Curtiss, and Langhoff, 1976). Unfortunately, the results of theoretical calculations differ considerably, mainly because of the large number of electrons to be correlated simultaneously. A correlation diagram of the states of O_3 in the ground state geometry (vertical) and the optimum C_{2v} geometries (adiabatic)* with states of O_2 and O is shown in Figure 2-1. The diagram should therefore be considered as illustrative rather than exact.

In spite of the differences in the results of theoretical calculations, some useful conclusions for ozone photochemistry can still be drawn from these calculations. Thus, on theoretical grounds there is a possibility that the a^3B_2 state may be bound by 0.2 eV (Hay and Dunning, 1977) or 0.4 eV (Wilson and Hopper, 1981) and may have an adiabatic excitation

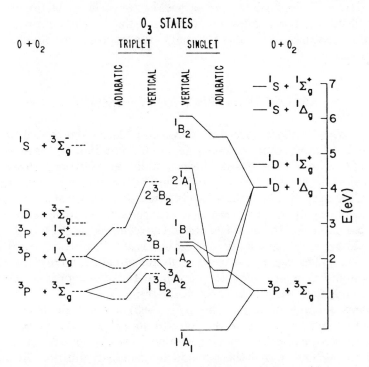

Figure 2-1. Correlation diagram of the states of O_3 at the ground state geometry (vertical) and optimum C_{2v} geometries (adiabatic) with the states of $O_2 + O$. *(From Hay, P. J., and T. H. Dunning, 1977, Geometries and energies of the excited states of O_3 from ab initio potential energy surfaces, J. Chem. Phys. 67:2299; copyright © 1977 by the American Institute of Physics.)*

*In a "vertical" representation of a transition, the excited state geometry of the corresponding potential energy surface is assumed to be the same as the ground state geometry. In an "adiabatic" representation, the transition is assumed to occur between minimum potential energy configurations of both the ground and the excited states. The adiabatic transition energy is always smaller than the vertical transition energy.

energy of either 0.9 eV or 0.7 eV, which are both less than the dissociation energy $D_e = 1.13$ eV of ground state ozone. Electron scattering experiments of Swanson and Cellota (1975) support this theoretical prediction. Moreover, the potential energy surface calculated by Wilson and Hopper did not show any barrier to the formation of this state from $O(^3P)$ and $O_2(^3\Sigma)$. It is probable that the $^3B_2$ state is formed in three-body O and O_2 recombination. An emission feature observed at 6.6 μm in the experiments of von Rosenberg and Trainor (1975) supports this conjecture. Excited $O_3(a^3B_2)$ is of potential importance in stratospheric and tropospheric photochemistry because the species is a possible source of nitrous oxide in the stratosphere (Prasad, 1981) and in the troposphere (Stedman and Shetter, 1982).

THE ABSORPTION SPECTRUM OF OZONE

With 18 valence electrons, the ozone molecule shows weak absorption in the photographic infrared as far as 1000 nm. Indeed, the longward limit of this absorption has not been established (Herzberg, 1944). Diffuse bands of O_3 in the photographic infrared are known as Wulf bands. According to Wulf (1930), these bands form a progression of ten members with a spacing of 567 cm^{-1}. Starting with the first member at 1000 nm, Wulf bands extend to about 660 nm and are centered at 827 nm (1.5 eV). Ozone absorption coefficients for these bands are ultraweak, about 0.0005 atm^{-1} cm^{-1}. The calculated band oscillator strength is only 1.9×10^{-7} (Wilson and Hopper, 1982). It is apparent that these properties make Wulf bands unimportant in photochemistry.

Chappuis bands of ozone (Chappuis, 1880) extend from 440-850 nm. They are most prominent in the 550-610 nm range, with the strongest bands at 573 nm and 602 nm. High-resolution studies by Humphrey and Badger (1947) show that these bands are genuinely diffuse. The occurrence of predissociation at these long wavelengths is not surprising, because the dissociation energy of ozone is only 1.1 eV. At long path lengths, Chappuis bands seem to merge into a very weak continuum (Wulf, 1930; Vigroux, 1953). The experimentally measured oscillator strength of Chappuis bands is 2×10^{-5} (Inn and Tanaka, 1953). Hay and Dunning (1977) attribute these bands to $X^1A_1 \rightarrow {}^1B_1$ transitions.

The Huggins bands of ozone begin at about 370 nm and extend to shorter wavelengths. First observed by Huggins (1890), the main bands of the system extend from 345-300 nm. The shortward end of the Huggins bands is overlapped by the shoulder of the Hartley bands. Although the Huggins bands are still diffuse, they are much sharper than the Chappuis bands. Most recent quantum number assignments, based on isotope effects, by Katayama (1979) suggest that the $(000)' \leftarrow (000)''$ transition occurs at 368.7 nm. Wilson and Hopper (1982) and Brand, Cross, and Hoy (1978) assign the bands to the $2^1A_1 \leftarrow X^1A_1$ transition. In contrast, Herzberg (1944), Simons et al. (1973) and Hay and Dunning (1977) assign these bands to the transition from the ground X^1A_1 state to the bound portions of the excited $^1B_2$ state. In this interpretation, the Huggins and Hartley bands have the same upper state.

The Hartley band of ozone (Hartley, 1881) consists of a broad continuum between 220 nm and 300 nm with a very high and almost symmetrical peak near 255 nm. The peak of the Hartley band at 255 nm has an absorption coefficient of 150 cm^{-1}, that is, a layer of 0.007 cm at atmospheric pressure absorbs 60% of the incident light. In particular, this is the reason that the upper atmospheric ozone layer of only 0.3 cm-atm completely absorbs all incident solar and stellar ultraviolet radiation in the 220-300 nm spectral region; it thereby provides a protective shield over the biosphere against lethal ultraviolet radiation. As implied earlier, the 1^1B_2 state is the upper state of the Hartley band.

Photolysis in the Huggins-Hartley ultraviolet region produces both O(^1D) and O(^3P). The O(^1D) atoms thus produced in the atmosphere are the most important drivers of atmospheric chemistry via the reactions with H_2O and N_2O. Considerable attention has therefore been focused on the yield of O(^1D) in ozone photolysis. Pertinent laboratory data (Davenport, 1980) appropriate for room temperature are shown in Figure 2-2. It should be noted that the quantum yield $\phi(O^1D)$ at $\lambda < 300$ nm is not unity, which is in agreement with the results of Amimoto et al. (1979), Brock and Watson (1980), Fairchild, Stone, and Lawrence (1978), and Sparks et al. (1980). These new results are, however, in sharp contrast with the widespread notion, held as recently as 1980, that $\phi(O^1D) = 1$ at $\lambda \le 300$ nm. At $\lambda > 300$ nm $\phi(O^1D)$ is a rapidly decreasing function of λ (Arnold, Comes, and Moortgat, 1977; Castellano and Schumacher, 1969; Jones and Wayne, 1970; Kajimoto and Cvetanovich, 1976; Kuis, Simonaitis, and Heicklen, 1975; Lin and DeMore, 1973/1974; Moortgat and Warneck, 1975; Philen, Watson, and Davis, 1977; Simonaitis et al., 1973). The precise position and the steepness of the decrease in $\phi(O^1D)$ is significantly dependent upon temperature (Moortgat, Kudszus, and Warneck, 1977). The long wavelength tail at $\lambda > 313$ nm appearing in Brock and Watson's (1980) data, however, does not appear in other measurements.

Because high absorption in the Hartley continuum efficiently removes light at wavelengths shorter than 300 nm in the stratosphere, actinic irradiance of photolytic light reaching the lower atmosphere (below 20 km) increases rapidly with increasing wavelength in the same wavelength domain ($\lambda > 300$ nm) in which $\phi(O^1D)$ declines steeply to negligible values. Consequently, O(^1D) production rates below 20 km are extremely sensitive to the wavelength variation of the quantum yield in the $300 \leq \lambda \leq 320$ nm range (Brock and Watson, 1980). Thus, for atmospheric chemistry in the lower stratosphere (below 20 km), the wavelength dependence of $\phi(O^1D)$ in the region 300 to 320 nm must be known precisely. The temperature dependence of $\phi(O^1D)$ in this region is also of critical importance. Considerable experimental and theoretical effort is therefore being devoted to the wavelength and temperature dependence of $\phi(O^1D)$ in the range $300 \leq \lambda \leq 320$ nm and $200 \leq T \leq 298$ K (Adler-Golden, Schweitzer, and Steinfeld, 1982). The temperature dependence of $\phi(O^1D)$ in the region where it rapidly decreases is caused by the contributions of internal rotational and vibrational energies to the total enery required to reach the $O(^1D)$ production threshold (Adler-Golden, Schweitzer, and Steinfeld, 1982; Moortgat, Kudszus, and Warneck, 1977; Zittel and Little, 1980). At present, the temperature and wavelength dependence of $\phi(O^1D)$ in the critical region 300 to 320 nm is thought to be best represented by an empirical expression developed by Moortgat and Kudzus (1978). Most

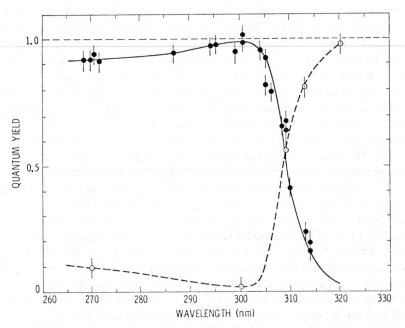

Figure 2-2. Room temperature values for atomic oxygen quantum yields: ●, (^1D) values based on measurement of $\phi(^1D) = 1.00$ at 300 nm; ☉, (^3P) absolute values. Lines are drawn primarily on the basis of $\phi(^3P)$ values. Vertical lines represent 90% uncertainty limits in determining the zero time intensity. *(From Davenport, J. E., 1980, Parameters of Ozone Photolysis as Functions of Temperature at 280–300 nm, Report FAA-EE-80-44R, SRI International, Menlo Park, Calif., p. 25.)*

recently, Adler-Golden, Schweitzer, and Steinfeld (1982) have provided an alternate expression which represents the long wavelength tail above 313 nm appearing in the Brock and Watson (1980) data. As pointed out earlier, this tail does not appear in other experimental data. Until the reality of the long wavelength tail is settled, the validity of the expression suggested by Adler-Golden, Schweitzer, and Steinfeld (1982) remains doubtful. However, the success of their expression at and below 313 nm suggests that vibrationally excited ozone does indeed play an important role in stratospheric ozone photolysis.

PHOTODISSOCIATION

As the "driver" of atmospheric photochemistry, photodissociation is of paramount importance to stratospheric ozone, both because of the formation of oxygen atoms from O_2 and because of the photolysis of other species that influence ozone abundance. The photodissociation rate of a molecule is calculated as the wavelength integral of the product of the solar intensity at a given point in the atmosphere and the photolysis cross section. However, determinations of photolysis rates generally make use of numerous approximation techniques to deal with complications associated with atmospheric absorption, including pressure broadening of spectral absorption lines and thermal excitation of rotational states, atmospheric scattering and surface albedo, and the mechanism of the dissociation process itself. If scattering and albedo are ignored, the general formula for computing photodissociation rates, J_i (s^{-1}), can be expressed as

$$J_i(z, \chi) = \int d\lambda \ F(\lambda)\sigma_i^P(\lambda, z) \exp[-\tau_\lambda(z, \chi)], \qquad (2\text{-}1)$$

where λ is wavelength, $\sigma_i^P(\lambda, z)$ is the photodissociation cross section at height z (accounting for possible temperature and pressure effects), and $\tau_\lambda(z, \chi)$ is the optical depth expressed as

$$\tau_\lambda(z, \chi) = \sum_j \int_z^\infty dz' \ n_j(z')\sigma_j^a(\lambda, z')\sec\chi. \qquad (2\text{-}2)$$

$F(\lambda)$ is the solar photon flux (photons cm^{-2} sec^{-1}) above the atmosphere. Here, $\sigma_j^a(\lambda, z)$ is the total absorption cross section of species j at wavelength λ, $n_j(z)$ is the concentration of species j at height z (molecules cm^{-3}), and χ is the solar zenith angle. Should χ exceed about 65 to 70°, $\sec \chi$ must be replaced by a Chapman function (e.g., Smith and Smith, 1972) that takes account of atmospheric sphericity. The rate of photodissociation of species i is

$$(dn_i/dt)_{photo} = -J_i(z)n_i(z). \qquad (2\text{-}3)$$

There are two types of dissociation mechanism of importance in the atmosphere. In direct dissociation, the molecule is excited directly from the ground state (characterized by the attractive potential energy curve X shown in Figure 2-3) to the repulsive potential energy curve B as illustrated by arrow a. The corresponding cross sections are usually smoothly varying with wavelength, displaying little if any structure. On the other hand, the dissociation might begin with excitation (arrow b) to another attractive state, A, which can either return to the ground state (arrow c) or transit to a low-lying repulsive state (C), which completes the dissociation. This process is called *predissociation*. As one would expect from quantum-mechanical arguments, transition rates for direct dissociation are usually much larger than for predissociation. Photodissociation of atmospheric constituents has been discussed in the literature by numerous investigators (e.g., Hudson et al., 1969; Hudson and

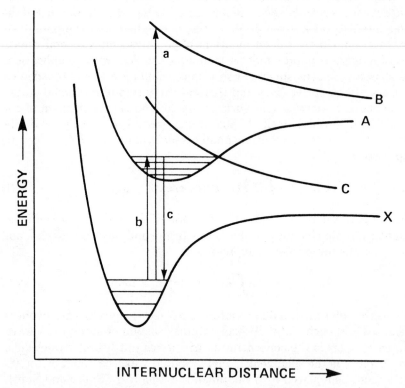

Figure 2-3. Potential energy diagram of a hypothetical diatomic molecule showing routes of direct and predissociation.

Mahle, 1972; Turco, 1975; Shimazaki and Helmle, 1977; Luther and Gelinas, 1976; Frederick and Hudson, 1979; Nicolet, 1979; Nicolet and Cieslik, 1980).

Atmospheric absorption of spectral lines is complicated by pressure broadening, which results from the distortion of molecular wave functions and thus energy states by collision with neighboring molecules. The degree of broadening (i.e., the "width" of the absorption line) is proportional to both the density and the temperature and thus to the pressure; hence the name "pressure broadening." Further complications ensue from the population of the various rotational states of the dissociating molecule. Occupation of rotational states above the ground state alters the positions of lines within the rotational-vibrational bands. The distribution of rotational state populations is essentially Maxwellian and thus temperature-dependent. This temperature dependence is significant in the absorption of solar radiation within the Schumann-Runge bands of O_2 lying in the range 175.9 nm to 195 nm (Hudson et al., 1969). For a more detailed discussion of pressure and temperature effects and spectral line absorption, see Chamberlain (1978).

For many photodissociation processes, multiple scattering and surface albedo cannot be ignored. In such cases, the appropriate form of the radiative transfer equation must be solved when computing the photolysis rate. The photolysis rate then takes the form

$$J_i(z, \lambda) = d\lambda' \, \sigma_i^p(\lambda, T)[F(\lambda'; \chi, z) + \int d\omega' \, I(\lambda; \omega', z)], \qquad (2\text{-}4)$$

where ω is an element of solid angle and $I(\lambda, \omega, z)$ is the intensity of the diffuse radiation (per unit solid angle), which is related to F through the radiative transfer equation (e.g., Chamberlain, 1978)

$$\mu \, dI(\lambda; \omega, z)/d\tau = I(\lambda; \omega, z) - (\tfrac{1}{4}\pi) \int d\omega' \, I(\lambda; \omega', z)p(\omega') \\ - \tfrac{1}{4} F(\lambda; \omega, z) \qquad (2\text{-}5)$$

and the definition

$$F(\lambda; \omega, z) \equiv F(\lambda)\exp(-\tau_\lambda/\mu_0)p(\omega; \chi, \phi), \qquad (2\text{-}6)$$

with the absorption cross section appearing in the element of optical depth (Eq. (2)) replaced by the sum of the absorption and scattering cross section, σ^a and σ^s respectively. Here $\mu_0 = \cos \chi$, ϕ is the solar azimuthal angle, $d\omega'$ is an element of solid angle, μ is the cosine of the angle between the photon momentum and the upward vertical, and $p(\omega')$ is the phase function dependent upon the scattering angle and defined such that

$$\tfrac{1}{4}\pi \int d\omega'\, p(\omega') = \sigma^a(\lambda)/[\sigma^a(\lambda) + \sigma^s(\lambda)]. \qquad (2\text{-}7)$$

Ratios of J (with atmospheric multiple scattering included) to J (for pure absorption only) have been computed by Luther and Gelinas (1976). For a number of molecules of atmospheric interest, scattering/albedo effects are quite large (e.g., for ozone), especially for dissociation into ground state oxygen atoms, and at low altitudes for photodissociation to produce $O(^1D)$ or $O(^3P)$. In calculating scattering/albedo effects, Luther and Gelinas computed radiation intensities at a given altitude for several discrete angles. Dissociation rates were then computed from the field intensities as just indicated. More recently, Meier, Anderson, and Nicolet (1982; also see Nicolet, Meyer, and Anderson, 1982) used an alternative approach explored earlier by other investigators (e.g., Yung, 1976). In this approach, they computed dissociation rates directly without an intermediate calculation of radiation intensity. Their results verified the earlier prediction of Luther and Gelinas (1976), but the computational method of Meier, Anderson, and Nicolet was faster and easier to use in photochemical models.

In performing calculations of photolysis rates, it would be ideal to carry out the integration over wavelength indicated in Eq. (2-1) over the complete spectral line contours as has been done by Hudson and Mahle (1972). However, since that approach makes heavy demands on computer resources, it is desirable to try to approximate the integral such that the final results are reasonably close (i.e., say, to within 20%) to that obtained with a "line-by-line" approach. The Elsasser band model (e.g., see Goody, 1964) is applicable where the lines of a band are regularly spaced, such as in the rotational-vibrational bands of diatomic molecules, if the wavelength interval is large compared to the vibrational structure. Then it can be shown that absorption in a limited wavelength interval, $\Delta\lambda$, by species i is

$$\tau(\lambda, \Delta\lambda) \simeq a_i(\lambda)[\int_z^\infty dz'\, n_i(z')]^{1/2}, \qquad (2\text{-}8)$$

where the parameter a_i is determined experimentally or theoretically for spectroscopic and line-broadening parameters. The effective optical depth increases only as the square root of the column depth of species i because of the "leakage" of radiation between spectral lines. As the absorption increases, the absorption lines become saturated and τ returns to a nearly linear dependence on the column depth (see Turco, 1975 for a more complete discussion). Shimazaki and Helmle (1977) have developed a particularly simple algorithm for use in computing solar radiation absorption in the Schumann–Runge bands, which is quite accurate. Briefly, they divide the spectrum into a small number on intervals of wavelength width

$\Delta\lambda_k$ and expand the integral of the absorption cross section in an exponential power series

$$\frac{1}{\Delta\lambda_k}\int_{\lambda_k}^{\lambda_{k+1}} d\lambda' \; \sigma_{O_2}(\lambda')\exp[-\tau_{O_2}(\lambda', z)] = \exp \sum_{n=0}^{7} \beta_n [ln\int_z^\infty dz' \; n_{O_2}(z')]^n, \quad (2\text{-}9)$$

where the expansion coefficients are obtained by fitting to the line-by-line results (Hudson and Mahle, 1972). Their results generally are within 50% of the line-by-line data with an enormous saving in computer time.

OZONE FORMATION AND DESTRUCTION IN THE UPPER ATMOSPHERE

Ozone is formed in the stratosphere by the photodissociation of molecular oxygen in the Herzberg bands and continuum (195-260 nm) and the Schumann-Runge bands (170-195 nm). The products are two ground state oxygen atoms

$$O_2 + h\nu \rightarrow 2 \; O(^3P). \quad (2\text{-}10)$$

Unfortunately, there is considerable uncertainty in the cross section of this key process, especially in the important 200-210 nm region. At these wavelengths Hasson and Nicholls' (1971) values (1.4×10^{-23} cm^2 and 1.1×10^{-23} cm^2 at 200 and 210 nm, respectively) differ by as much as 40% from Shardanand and Prasad Rao's (1977) values of 1×10^{-23} cm^2 and 7.7×10^{-24} cm^2 respectively. Frederick and Mentall (1982) and Herman and Mentall (1982) have critically investigated this uncertain situation in the light of absorption measurements made during a balloon flight. These investigations indicate that the lower values reported by Shardanand and Prasad Rao (1977) are more likely to be closer to the true value. Further laboratory measurements are needed. Globally averaged photodissociation rates of molecular oxygen and ozone computed by means of the method of Turco and Whitten (1977) are shown in Figure 2-4.

Sydney Chapman (1930) proposed the ozone formation theory that still forms the basis for all the work in this area. Briefly, the oxygen atoms formed by the photolysis of O_2 rapidly associate with the very abundant oxygen molecules by the three-body association reaction

$$O + O_2 + M \rightarrow O_3 + M. \quad (2\text{-}11)$$

In Chapman's scheme, destruction of ozone, that is, its conversion back to O_2, occurs via the reaction

$$O + O_3 \rightarrow 2\,O_2. \tag{2-12}$$

It was realized long ago that reaction (2-12) is too slow to account for the observed abundance of stratospheric ozone. It was therefore suggested (Hampson, 1964; Hunt, 1966) that the dominant loss mechanism occurs via a catalytic sequence of reactions involving OH and HO_2 radicals. The resulting loss rates still proved to be too small, leading Crutzen (1970) to suggest that analogous reactions with nitrogen oxides, NO and NO_2, would provide the necessary loss mechanism. Finally, Stolarski and Cicerone (1974) proposed that ozone can also be catalytically destroyed by reactions with chlorine and chlorine monoxide (ClO) radicals. Each of these mechanisms as well as the sources and sinks of reactive catalysts will be discussed in the following sections.

However, before going on to discuss the catalytic reactions, it will prove useful to define the term "odd-oxygen." Odd-oxygen refers to the family of species that includes atomic oxygen, ozone, and those other species from which O is readily released (e.g., nitrogen dioxide and chlorine monoxide).

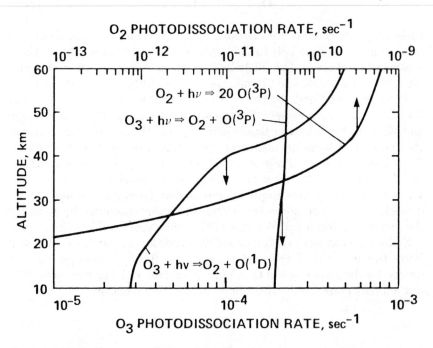

Figure 2-4. Globally averaged photodissociation rates of O_2 and O_3 computed by the method of Turco and Whitten (1977).

THE PHOTOCHEMISTRY OF ODD-HYDROGEN

Odd-hydrogen (HO_x) is defined as those species that contain a single hydrogen atom (e.g., the radicals H, OH, and HO_2), and it often includes those stable compounds like HNO_3 and H_2O_2 that are easily dissociated into odd-hydrogen radicals. The HO_x family was first studied in its atmospheric context by Bates and Nicolet (1950). All recent advances stem from their work. The photochemical cycle of odd-hydrogen species in the present stratosphere is shown in Figure 2-5. It involves three groups of species: source species (H_2O, H_2, and CH_4), reactive radical species (OH and HO_2), and the reservoirs (H_2O_2, HNO_3, HO_2NO_2, and HOCl).

In the stratosphere, odd-hydrogen is formed primarily by the reaction of metastable $O(^1D)$ atoms resulting from ozone photolysis:

$$O_3 + h\nu \rightarrow O_2(^1\Delta_g) + O(^1D) \qquad (\lambda < 310\ nm - Hartley\ bands) \qquad (2\text{-}13)$$

with water vapor or with molecular hydrogen:

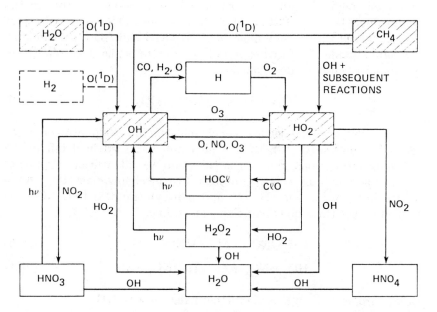

Figure 2-5. The photochemical cycle of odd-hydrogen in the present stratosphere.

$$O(^1D) + H_2O \rightarrow 2\ OH \qquad (2\text{-}14)$$

$$O(^1D) + H_2 \rightarrow OH + H. \qquad (2\text{-}15)$$

A secondary source is the reaction of methane with already existing hydroxyl,

$$OH + CH_4 \rightarrow CH_3 + H_2O, \qquad (2\text{-}16)$$

where the CH_3 eventually decomposes into hydrogen radicals, thus constituting a net source of odd-hydrogen. This point will become evident at a later stage.

The hydrogen atoms formed by reaction (2-15) react rapidly with the very abundant molecular oxygen to form HO_2:

$$H + O_2 + M \rightarrow HO_2 + M. \qquad (2\text{-}17)$$

All the odd-hydrogen radicals can recombine with one another to form H_2 and H_2O along several paths, the most important of which is initiated by the radical disproportionation

$$HO_2 + HO_2 \rightarrow H_2O_2 + O_2. \qquad (2\text{-}18)$$

Hydrogen peroxide can be found in relatively large amounts in the stratosphere because of its small photolysis rate and low reactivity. This species (along with related molecules CH_3OOH and HO_2NO_2) can limit the total abundance of hydrogen radicals in stratospheric air via hydrogen abstraction reactions such as

$$OH + H_2O_2 \rightarrow H_2O + HO_2. \qquad (2\text{-}19)$$

Odd-hydrogen is coupled to other chemical families: odd-nitrogen (NO_y) and odd-chlorine (Cl_x), both of which are discussed in detail in the following two sections. However, at this point we will mention coupling to odd-nitrogen because it is a significant loss mechanism for HO_x. In the atmosphere, nitric acid (HNO_3) and peroxynitric acid (HO_2NO_2) are formed by the three-body association reactions

$$OH + NO_2 + M \rightarrow HNO_3 + M \qquad (2\text{-}20)$$

$$HO_2 + NO_2 + M \rightarrow HO_2NO_2 + M. \qquad (2\text{-}21)$$

Odd-hydrogen is then lost via further reaction with OH

$$OH + HNO_3 \rightarrow H_2O + NO_3 \qquad (2\text{-}22)$$

(Nelson, Marinelli, and Johnston, 1981; Wine et al., 1981) and

$$OH + HO_2NO_2 \rightarrow H_2O + O_2 + NO_2 \qquad (2\text{-}23)$$

(Barnes et al., 1981; Trevor, Black, and Barker, 1982). Both reactions (2-22) and (2-23) are now known to be significant loss mechanisms for stratospheric HO_x.

A second important coupling between HO_x and odd-nitrogen involves the reaction

$$NO + HO_2 \rightarrow NO_2 + OH, \qquad (2\text{-}24)$$

which significantly affects the $OH : HO_2$ concentration ratio in the lower stratosphere. The fast kinetic rate coefficient for reaction (2-24), which was originally measured by Howard and Evenson (1977), led to predictions of substantially larger OH abundance in the lower stratosphere. However, the realization that the HO_x loss rates due to reactions (2-22) and (2-23) are very large has decreased the perceived importance of reaction (2-24).

Although not important to HO_x abundance in the ambient stratosphere, we must mention the reactions

$$OH + SO_2 + M \rightarrow HSO_3 + M \qquad (2\text{-}25)$$

$$HSO_3 + OH \rightarrow H_2O + SO_3 \qquad (2\text{-}26)$$

that may significantly reduce OH (and HO_2) abundances after the injection of SO_2 into the stratosphere by a volcanic eruption.

Although reaction (2-16) appears to lead to the destruction of odd-hydrogen, it does not do so, because the methane radical CH_3 is itself a source of odd-hydrogen. In general, reaction (2-16) produces more odd-hydrogen than it destroys. Figure 2-6 gives a schematic representation of the methane oxidation sequence in the stratosphere. Interestingly, methane decomposition in the atmosphere leads to ozone production via a photochemical mechanism similar to that occurring in polluted urban atmospheres (smog). The yields of odd-hydrogen and odd-oxygen production and loss at each reaction step and their net production for the entire reaction sequence are also indicated in Figure 2-6.

One component of the methane smog reaction mechanism involves the methyl radicals produced by reaction (2-16). These radicals react rapidly to form a peroxy compound somewhat analogous to HO_2:

$$CH_3 + O_2 + M \rightarrow CH_3O_2 + M, \qquad (2\text{-}27)$$

which then reacts with NO:

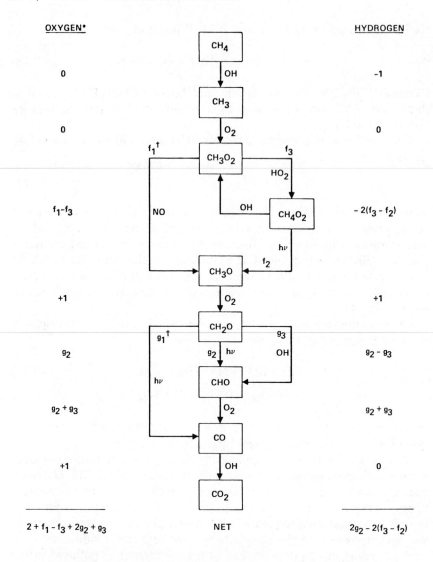

OXYGEN*

0

0

$f_1 - f_3$

+1

g_2

$g_2 + g_3$

+1

$2 + f_1 - f_3 + 2g_2 + g_3$

HYDROGEN

-1

0

$-2(f_3 - f_2)$

+1

$g_2 - g_3$

$g_2 + g_3$

0

$2g_2 - 2(f_3 - f_2)$

NET

*FRACTIONAL NUMBERS OF ATOMS GENERATED FOR EACH METHANE MOLECULE OXIDIZED. TO OBTAIN THESE FRACTIONS, UNIT EFFICIENCY IS ASSUMED AT EACH STEP OF THE RELATED CHEMICAL CHAIN.

$$H \xrightarrow{O_2} HO_2 \xrightarrow{NO} NO_2 \xrightarrow{h\nu} O \xrightarrow{O_2} O_3$$

†THE f's AND g's ARE BRANCHING RATIOS; BRANCH f_3, HOWEVER, IS NORMALIZED RELATIVE TO $f_1 + f_2 = 1$.

Figure 2-6. The methane oxidation sequence

$$CH_3O_2 + NO \rightarrow CH_3O + NO_2. \tag{2-28}$$

Reaction (2-28), the analog of reaction (2-24), followed by the photolysis of NO_2

$$NO_2 + h\nu \rightarrow NO + O \tag{2-29}$$

and reaction (2-11) generates ozone.

The concentration of CH_3O_2 is limited mainly by reaction (2-28). However, CH_3O_2 can also react with HO_2

$$CH_3O_2 + HO_2 \rightarrow CH_3OOH + O_2 \tag{2-30}$$

to form the hydrocarbon analog of hydrogen peroxide. The CH_3OOH molecules that are formed may also be photolyzed into CH_3O and OH (products analogous to H_2O_2 photodissociation fragments), short-circuiting ozone production via reaction (2-28), or they may react with OH, reducing odd-hydrogen concentrations:

$$CH_3OOH + OH \rightarrow CH_3O_2 + H_2O. \tag{2-31}$$

However, using present knowledge of the kinetic data, it is obvious that such CH_3OOH effects are small.

Another important factor in methane oxidation chemistry is the branching ratio for the production of radicals (H and CHO) and molecules (H_2 and CO) in the photolysis of formaldehyde. The quantum yield data of Moortgat el al. (1978) favor the formation of HO_x radicals at the expense of H_2 and CO.

In the present context, the principal significance of odd-hydrogen is the catalytic destruction of ozone both directly and through its effects on odd-nitrogen and odd-chlorine. In the lower stratosphere, hydrogen radicals catalytically destroy ozone molecules through the reaction sequence

$$OH + O_3 \rightarrow HO_2 + O_2 \tag{2-32}$$

$$\underline{HO_2 + O_3 \rightarrow OH + 2O_2} \tag{2-33}$$
$$2\,O_3 \rightarrow 3\,O_2 \qquad \text{(overall reaction)}.$$

At higher altitudes, odd-oxygen is consumed by the cycles

$$OH + O \rightarrow H + O_2 \tag{2-34}$$

$$H + O_2 + M \rightarrow HO_2 + M \tag{2-35}$$

$$\underline{HO_2 + O \rightarrow OH + O_2} \tag{2-36}$$
$$O + O \rightarrow O_2 \qquad \text{(overall reaction)},$$

$$OH + O \rightarrow H + O_2 \tag{2-37}$$

$$\frac{H + O_3 \rightarrow OH + O_2}{O + O_3 \rightarrow 2\,O_2} \tag{2-38}$$
$$\text{(overall reaction)},$$

and

$$HO_2 + O \rightarrow OH + O_2 \tag{2-39}$$

$$\frac{OH + O_3 \rightarrow HO_2 + O_2}{O + O_3 \rightarrow 2\,O_2} \tag{2-40}$$
$$\text{(overall reaction)}.$$

Interestingly, reaction (2-24) followed by reactions (2-29) and (2-11) inter-feres with HO_x ozone catalysis by competing with catalytic steps (2-32) and (2-36); thus the reaction of NO with HO_2 acts to decrease ozone consumption by hydrogen compounds. Odd-oxygen loss rates due to reactions (2-12), (2-33), (2-34), and (2-36) are shown in Figure 2-7 as a function of altitude. Evidently, reaction (2-34) becomes significant only as one approaches the mesosphere, while reaction (2-33) is an important loss process only in the lower stratosphere; nowhere does reaction (2-12) dominate the odd-oxygen loss processes. Some of the revisions of our knowledge of HO_x chemistry relative to stratospheric ozone brought

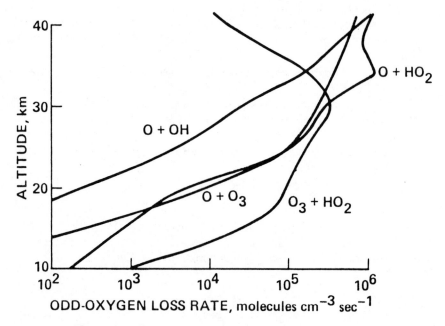

Figure 2-7. Computed odd-oxygen loss rates due to various HO_x reactions and to $O + O_3 \rightarrow 2\,O_2$.

about by recent rate coefficient measurements have been discussed by Crutzen and Howard (1978), Turco et al. (1978), Whitten et al. (1981), and in a World Meteorological Organization report (1981).

THE PHOTOCHEMISTRY OF ODD-NITROGEN

Odd-nitrogen (NO_y) is defined as the family of species that contains one or more free nitrogen atoms (e.g., N, NO, NO_2, NO_3, N_2O_5, HNO_3, HO_2NO_2, and $ClONO_2$); NO_y is not easily produced in the atmosphere because nitrogen molecules are notoriously difficult to dissociate. Once formed, however, NO_y is not easily removed; stratospheric loss processes are so inefficient that NO_y must be transported to the mesosphere or ground before it can be destroyed efficiently. A subset of the odd-nitrogen family composed of NO, NO_2, and NO_3, and referred to as NO_x, is frequently distinguished from the other species because of its rapid reaction with ozone and atomic oxygen.

The cycle of reactive NO_x species is shown in Figure 2-8. Nitrous oxide

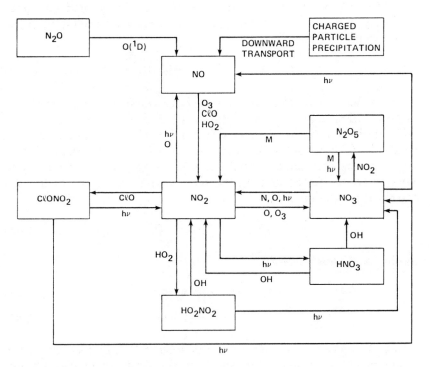

Figure 2-8. The photochemical cycle of odd-nitrogen in the present stratosphere.

(N_2O) is by far the most important source of NO_x in the stratosphere (McElroy and McConnell, 1971; Nicolet, 1971) through the reaction

$$N_2O + O(^1D) \rightarrow 2\ NO. \tag{2-41}$$

This source species is released into the atmosphere as a result of natural microbiological activities and anthropological activities such as burning of fossil fuels (Hahn, 1979; Hahn and Junge, 1977; Liu et al., 1977; Logan et al., 1978; McElroy et al., 1976). Despite the vast research effort carried out, there is little agreement on the magnitude of the natural sources and sinks of nitrous oxide or indeed the nitrogen fixation cycle in general. Hence, likely perturbations of atmospheric N_2O by anthropogenic activity (see Chapter 5) are still highly speculative. Recently, yet another new dimension has been added to this already complex problem with the discovery of nitrous oxide production from the reaction

$$\begin{array}{ll} \rightarrow N_2O + H & \text{(2-42a)} \\ OH(A^2\Pi) + N_2 & \\ \rightarrow OH(X^2\Pi) + N_2 & \text{(2-42b)} \end{array}$$

(Prasad and Zipf, 1982). Excited $OH(A^2\Pi)$ radicals are produced quite efficiently in the upper stratosphere by solar resonant excitation of the ambient OH radicals

$$OH(X^2\Pi) + h\nu \rightarrow OH(A^2\Pi). \tag{2-43}$$

The new source therefore appears to have considerable potential for forming odd-nitrogen (Prasad and Zipf, 1982).

Global distribution of nitrous oxide is quite uniform (Singh, Salas, and Shigeshi, 1979; Weiss, 1981), demonstrating that its tropospheric lifetime must be very large or that the sources and sinks must be quite homogeneously distributed. In the stratosphere, nitrous oxide may be destroyed by photolysis

$$N_2O + h\nu \rightarrow N_2 + O \tag{2-44}$$

and by reaction with $O(^1D)$

$$N_2O + O(^1D) \rightarrow N_2 + O_2 \tag{2-45}$$

$$\rightarrow 2\ NO. \tag{2-41}$$

In addition to the formation of stratospheric NO_y by reaction (2-41), odd-nitrogen by be injected directly into the stratosphere by high altitude aircraft (e.g., Oliver et al., 1977) and by large nuclear explosions (Bauer and Gilmore, 1975; Chang, Duewer, and Wuebbles, 1979; Johnston, Whitten, and Birks, 1973; Whitten, Borucki, and Turco, 1975). Finally, NO_y is formed in both the stratosphere and the mesosphere by charged-particle bombardment, particularly at high latitudes, and by large entry bodies such as meteors and space craft. Ionization by galactic cosmic ray particles leads to a net equivalent production of NO molecules of $\sim 5 \times 10^7$ cm^{-2} s^{-1} at high latitudes (Nicolet, 1975). Odd-nitrogen of galactic cosmic-ray origin is a negligible source insofar as ozone chemistry is concerned. There is a slight variation over the solar cycle due to solar wind modulation of the cosmic ray spectrum. In the mesosphere, NO_y is formed by ionization resulting from relativistic electron precipitation (Thorne, 1980) and solar proton bombardment (Crutzen, Isaksen, and Reid, 1975; Heath, Krueger, and Crutzen, 1977; Reagan et al., 1981; Reid, McAfee, and Crutzen, 1978). Odd-nitrogen production during exceptionally large solar proton events, such as that of August 1972, appears to be sufficient to cause an observable decrease in stratospheric ozone (Crutzen, Isaksen, and Reid, 1975). It is possible that NO_y production from precipitating charged particles may also be significant for stratospheric photochemistry on a long-term basis. Meteor and meteoroid bombardment (Henderson-Sellers, 1977; Park, 1978; Park and Menees, 1978) and space-craft reentry (Park, 1976, 1980; Whitten et al., 1982) are capable of generating large amounts of NO_y locally along their trajectories. The long-term influence of these odd-nitrogen sources is doubtful, however.

Tropospheric NO_y is produced both by lightning (e.g., Chameides et al., 1977; Levine et al., 1979; Tuck, 1976) and by combustion. However, it is removed from the atmosphere before it can be transported into the stratosphere.

Once formed in the stratosphere or transported into that region from the high latitude mesosphere, NO_y species are quite stable. In fact they must be transported either upward into the low latitude mesosphere or downward into the troposphere to be destroyed. Tropospheric removal results from the high solubility of nitric acid and nitric oxide in water—it is entrained in water droplets and removed by precipitation processes. In the mesosphere, odd-nitrogen destruction occurs via the two-step process

$$NO + h\nu \rightarrow N + O \qquad (2\text{-}46)$$

$$N + NO \rightarrow N_2 + O \qquad (2\text{-}47)$$

where process (2-46) occurs via predissociation in the NO δ bands at 181.3-183.5, 189.4, and 191.6 nm wavelength ranges. Calculation of the NO photodissociation rate is complicated by the rotational fine structure in both NO and NO_2 cross sections. The most recent calculations due to

Frederick and Hudson (1979), Nicolet (1979), and Nicolet and Cieslik (1980) are shown in Figure 2-9. Using the earlier NO oscillator strength data, which overestimated the photolysis rate by ~3.6, Duewer, Wuebbles, and Chang (1977) estimated that NO_y abundance in the upper stratosphere should be reduced by a factor of two, but this value must be reduced in light of the newer photolysis data.

Nitrogen oxides are closely coupled to the odd-oxygen system. Indeed, as we saw in a previous section, NO_2 must be considered as a carrier of odd-oxygen. We will return to the odd-oxygen coupling, but first we must investigate coupling to odd-hydrogen and to chlorine.

We have already mentioned in the preceding section the three-body association reaction of NO_2 with OH to form nitric acid. This reaction constitutes a very effective removal mechanism for NO_x ($NO + NO_2 + NO_3$), storing the odd-nitrogen in the form of HNO_3; of course, the total odd-nitrogen (NO_y) is unaffected, since there is no way to convert stratospheric nitric acid vapor into N_2. In principle, nitrous acid can also serve as an NO_y reservoir, but its rapid photolysis (Cox and Derwent, 1976; Zafonte, Rieger, and Holmes, 1975) renders it insignificant in that role. However, there are other species, peroxynitric acid (HO_2NO_2), chlorine nitrate ($ClONO_2$), and nitrogen pentoxide (N_2O_5), all formed by three-body association reactions

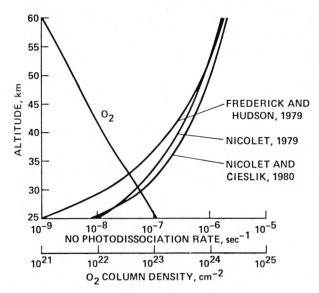

Figure 2-9. Computed nitric oxide photolysis rates due to various investigators and the mean O_2 column density above the indicated latitude.

$$HO_2 + NO_2 + M \rightarrow HO_2NO_2 + M \qquad (2\text{-}48)$$

$$ClO + NO_2 + M \rightarrow ClONO_2 + M \qquad (2\text{-}49)$$

$$NO_2 + NO_3 + M \rightarrow N_2O_5 + M, \qquad (2\text{-}50)$$

which have some significance as reservoirs, although less than nitric acid. The corresponding photodissociation cross sections for HNO_3, HO_2NO_2, $ClONO_2$, and N_2O_5 due to Hudson and Kieffer (1975) (also see Johnston and Graham, 1974; Molina and Molina, 1979; Molina and Molina, 1980; and Graham and Johnston, 1978 respectively), lead to the mid-latitude computed photolysis rates shown in Figure 2-10. While reactions (2-48) and (2-49) couple the NO_y system to the HO_x and Cl_x systems, reaction (2-50) is a purely self-coupling mechanism. Of the three species, HO_2NO_2 is the most significant (primarily because of its role in the destruction of HO_x discussed earlier) and N_2O_5 the least because it is rapidly dissociated by sunlight—except perhaps in the polar night and because stratospheric nitrogen trioxide (NO_3), formed by the (rather slow) reaction

$$NO_2 + O_3 \rightarrow NO_3 + O_2 \qquad (2\text{-}51)$$

is not very abundant. It has been found from two-dimensional model studies (unpublished) that N_2O_5 is of marginal significance as an NO_y reservoir except during the polar night.

The synergistic relationship between NO_y and Cl_x has some interesting consequences, which will be more fully explored in the following section.

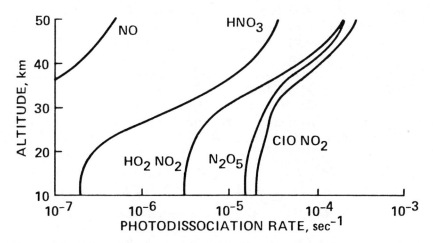

Figure 2-10. Computed globally averaged photolysis rates of HNO_3, HO_2NO_2, $ClONO_2$, and N_2O_5 by the method of Turco and Whitten (1977).

Suffice it to say at this point that reaction (2-49), which makes $ClONO_2$, a chlorine as well as an NO_y reservoir, is partially offset by the reaction

$$NO + ClO \rightarrow Cl + NO_2. \qquad (2\text{-}52)$$

In other words, an increase in the rate of reaction (2-52) caused by the addition of NO_y to the atmosphere, leads to a much smaller increase in the rate of formation of $ClONO_2$ than would be the case if reaction (2-52) did not occur.

As was shown in the preceding section, the coupling of NO_y to HO_x leads to destruction of the latter. The reaction sequence is catalytic in nature since no NO_y is lost in the process. For example,

$$OH + HNO_3 \rightarrow H_2O + NO_3 \qquad (2\text{-}53)$$

$$NO_3 + h\nu \rightarrow NO_2 + O \qquad (2\text{-}54)$$

$$\underline{NO_2 + OH + M \rightarrow HNO_3 + M} \qquad (2\text{-}55)$$
$$OH + OH \rightarrow H_2O + O \quad \text{(overall reaction)}$$

and

$$OH + HO_2NO_2 \rightarrow H_2O + O_2 + NO_2 \qquad (2\text{-}23)$$

$$\underline{NO_2 + HO_2 + M \rightarrow HO_2NO_2 + M} \qquad (2\text{-}56)$$
$$OH + HO_2 \rightarrow H_2O + O_2 \quad \text{(overall reaction).}$$

Of course, reactions (2-23) and (2-56) also destroy some odd-oxygen since HO_2 is a potential odd-oxygen radical. However, the cycle is an insignificant sink for odd-oxygen compared to other processes. It should be pointed out that photolysis of NO_3 also leads to the production of NO and O_2, but photoprocess (54) accounts for 90 percent of the decomposition (Magnotta and Johnston, 1980).

As was mentioned in a previous section, Crutzen (1970) first proposed NO_x reactions with odd-oxygen as an important mechanism for ozone loss. The reaction steps are

$$NO + O_3 \rightarrow NO_2 + O_2 \qquad (2\text{-}57)$$

$$\underline{NO_2 + O \rightarrow NO + O_2} \qquad (2\text{-}58)$$
$$O + O_3 \rightarrow 2\,O_2 \quad \text{(overall reaction),}$$

of which reaction (2-58) is the rate-controlling step. Reaction (2-57) can be thought of as a transfer of oxygen atoms from oxygen molecules to nitric oxide molecules. Reaction (2-58) then recombines two oxygen atoms with the NO molecule acting as an (originally bound) third body required to carry off the energy released by the reaction. Figure 2-11 shows the rate of reaction (2-58) as a function of altitude.

THE PHOTOCHEMISTRY OF ODD-CHLORINE

Paralleling the definition of odd-hydrogen and odd-nitrogen species, the term "odd-chlorine" (Cl_x) includes Cl, ClO, HCl, $ClONO_2$, and HOCl. They are important because reactive Cl and ClO catalytically consume O_3 and O. Figure 2-12 shows the various source, sink, and reactive radical species involved in atmospheric chlorine chemistry and their relationship to each other.

With the exception of methyl chloride (CH_3Cl), which is thought to come from the oceans and from the burning of vegetation (Lovelock, 1975; Singh et al., 1979; Watson, Lovelock, and Stedman, 1980), the presence of all other source species in the atmosphere is due to leakage in the course of their industrial use. For example, chlorofluoromethanes (CF_2Cl_2 and $CFCl_3$) are released into the atmosphere from spray cans and

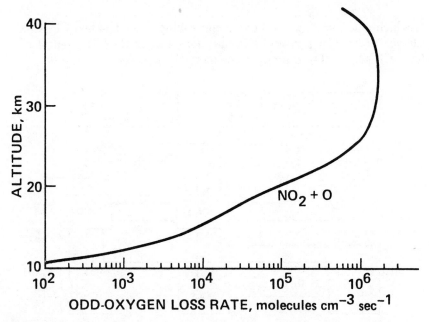

Figure 2-11. Computed loss rate of odd-oxygen due to reaction (58) as a function of altitude.

refrigeration devices. With the growth of industrial activities, the atmospheric abundance of these species has been increasing. Although the rate of this growth is not known with precision, workable estimates are available from various studies such as those reviewed by Cicerone (1981) and Jesson (1980). Additional details are presented in Chapter 5. Upon their release into the troposphere near the surface of the earth, these gases are transported upward into the stratosphere. Methyl chloride and methyl chloroform can be removed from the atmosphere by hydrogen abstraction reactions with the hydroxyl (OH) radical, so that only a fraction of the initial release of these substances is able to reach the stratosphere. In contrast, no significant tropospheric sink has been identified for non-hydrogen-bearing chlorofluoromethanes. Almost all the chlorofluoromethane releases are therefore able to find their way into the stratosphere. Once in the stratosphere, all source gases generate atomic chlorine by photodissociation. Atomic chlorine generated in this manner initiates the most important ozone destruction cycle attributable to the chlorine family:

$$Cl + O_3 \rightarrow ClO + O_2 \tag{2-59}$$

$$\frac{ClO + O \rightarrow Cl + O_2}{O_3 + O \rightarrow Cl + 2\,O_2} \tag{2-60}$$
(overall reaction),

of which reaction (2-60) is the rate-controlling step. The fast $Cl-ClO-Cl$ cycle started by a given Cl atom is halted when the latter reacts with CH_4, CH_2O, or HO_2 to produce the inert HCl species

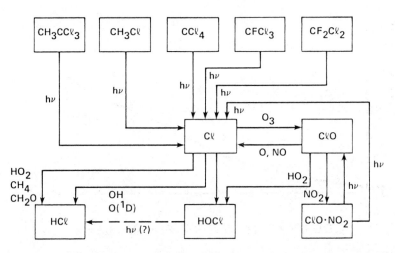

Figure 2-12. The photochemical cycle of odd-chlorine in the present stratosphere.

$$Cl + CH_4 \rightarrow HCl + CH_3 \qquad (2\text{-}61)$$

$$Cl + CH_2O \rightarrow HCl + CHO \qquad (2\text{-}62)$$

$$Cl + HO_2 \rightarrow HCl + O_2. \qquad (2\text{-}63)$$

According to current (1983) understanding, the source strength of ClO from Cl via reaction (2-59) is estimated to be 3×10^6 cm^{-3} s^{-1} at 30 km altitude, local noon and midlatitude. In contrast, the source strength of HCl from Cl via reactions (2-61) through (2-63) is only about 1×10^3 cm^{-3} s^{-1}. In other words, a chlorine atom completes many cycles which destroy ozone and oxygen before being trapped in the inert HCl reservoir. This action explains the importance of the cycle composed of reactions (2-59) and (2-60). This cycle is also terminated by reactions of ClO that lead to HOCl and ClONO$_2$ formation:

$$ClO + NO_2 + M \rightarrow ClONO_2 + M \qquad (2\text{-}64)$$

$$ClO + HO_2 \rightarrow HOCl + O_2. \qquad (2\text{-}65)$$

For example, at 30 km altitude, local noon and midlatitude, the rate at which ClONO$_2$ and HOCl are converted into reactive Cl and ClO are about 8×10^3 and 2.5×10^3 cm^{-3} s^{-1} respectively. These are significantly larger than the rate (1.5×10^3 cm^{-3} s^{-1}) at which HCl is converted into Cl via the reactions

$$HCl + OH \rightarrow H_2O + Cl \qquad (2\text{-}66)$$

$$HCl + O(^1D) \rightarrow OH + Cl \qquad (2\text{-}67)$$

Chlorine nitrate (ClONO$_2$) and hypochlorous acid (HOCl) are therefore less important reservoir species. In fact, because of their relatively larger photodissociation rates, ClONO$_2$ and HOCl are capable of driving the following two additional odd-oxygen destruction cycles

$$Cl + O_3 \rightarrow ClO + O_2 \qquad (2\text{-}59)$$

$$ClO + HO_2 \rightarrow HOCl + O_2 \qquad (2\text{-}65)$$

$$OH + O_3 \rightarrow HO_2 + O_2 \qquad (2\text{-}40)$$

$$\underline{HOCl + h\nu \rightarrow Cl + OH} \qquad (2\text{-}68)$$

$$2\,O_3 \rightarrow 3\,O_2 \qquad \text{(overall reaction)}$$

$$Cl + O_3 \rightarrow ClO + O_3 \tag{2-59}$$

$$ClO + NO_2 + M \rightarrow ClONO_2 + M \tag{2-64}$$

$$NO + O_3 \rightarrow NO_2 + O_2 \tag{2-57}$$

$$ClONO_2 + h\nu \rightarrow NO_3 + Cl \tag{2-69}$$

$$NO_3 + h\nu \rightarrow NO + O_2 \tag{2-70}$$

$$\overline{2\,O_3 \rightarrow 3\,O_2} \qquad \text{(overall reaction).}$$

The contribution of the cycle represented by reactions (2-40), (2-59), (2-65), and (2-68) to ozone destruction is about 25% at 25 km altitude. It decreases rapidly with increasing altitude to less than 1% at 40 km. Between 15 km and 25 km, all three cycles are competitive. At the lower altitudes, however, odd-oxygen destruction is dominated by catalytic cycles involving nitrogen. Furthermore, because the production rate of NO_3 is not controlled by the photolysis rate of chlorine nitrate, ClO is the only member of the chlorine family involved in a rate-limiting odd-oxygen step. As has been remarked, that the mode of photolysis of NO_3 (2-54) is considerably stronger (by a factor of about 10) than the mode (2-70) (Graham and Johnston, 1978; Magnotta and Johnston, 1980).

Accordingly, considerable importance is attached to the partitioning of chlorine into the rate-limiting radical ClO. In the present stratosphere, HCl is believed to be the major inorganic chlorine compound, and its partitioning into ClO is governed by six major reactions

$$HCl + OH \rightarrow H_2O + Cl \tag{2-66}$$

$$Cl + CH_4 \rightarrow HCl + CH_3 \tag{2-61}$$

$$Cl + HO_2 \rightarrow HCl + O_2 \tag{2-63}$$

$$Cl + O_3 \rightarrow ClO + O_2 \tag{2-59}$$

$$ClO + NO \rightarrow Cl + NO_2 \tag{2-71}$$

$$ClO + O \rightarrow Cl + O_2 \tag{2-60}$$

Three of these reactions exchange Cl with HCl and the other three exchange Cl with ClO. At altitudes above 35 km, the ratio between the concentrations of ClO and HCl, that is, [ClO]/[HCl] (where the brackets denote the concentration of the species), can be approximated by

$$[ClO]/[HCl] \simeq k_{66}[OH]k_{59}[O_3]/k_{61}[CH_4]k_{73}[NO]. \tag{2-72}$$

In this region, [ClO] is very sensitive to [OH] because [CH_4] is affected by [OH] in a synergistic manner. Below 35 km, however, the ClO concentration is also influenced by the formation of $ClONO_2$. But the key point is that the dependence of [ClO] on [OH] is quite strong in this region.

Altitude profiles of the observed and calculated HCl/ClO concentration ratio are shown in Figure 2-13. The profiles represented by the solid curves are based upon the set of rate coefficients recommended by DeMore et al. (1982), while the profiles represented by the broken lines are based upon older chemical kinetic data in which the reaction

$$OH + HO_2NO_2 \rightarrow H_2O + O_2 + NO_2 \qquad (2\text{-}23)$$

was relatively unimportant. It is apparent that with the new kinetic data, the agreement between theory and observation is quite good. The NASA-Ames two-dimensional stratospheric photochemical model (Borucki et al., 1980; Whitten et al. 1977) was used for the computations.

It has been suggested by Prasad (1980) that the reaction

$$ClO + O_2 + M \underset{\leftarrow}{\overset{\rightarrow}{\rightleftharpoons}} OClOO + M \qquad (2\text{-}73)$$

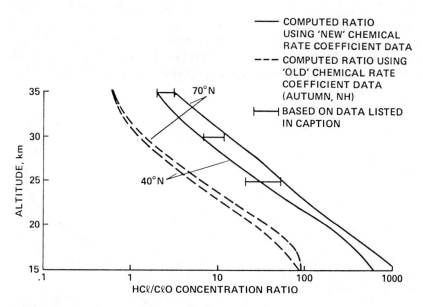

Figure 2-13. Computed profiles of HCl/ClO concentration ratios of 40° N and 70° N latitude in autumn (northern hemisphere) using rate coefficients evaluated by DeMore et al. (1982) (solid curves) and rate coefficients used by Whitten et al. (1981) (broken curves); HCl distributions were taken from Ackerman et al. (1976), Buijs et al. (1980), Farmer, Raper, and Norton (1976), Raper et al. (1977), Williams et al. (1976), and Zander (1981); ClO distributions are taken from Weinstock, Phillips, and Anderson (1981).

may be of some importance in the altitude range 25-35 km. The possible existence and chemistry of OClOO was originally inferred from an analysis of the suppression of chlorine catalyzed ozone dissociation by molecular oxygen (Prasad and Adams, 1980; Wongdontri-Stuper et al., 1979). A recent independent experiment by Zellner and Handwerk (1982) appears to lend additional support to the existence of OClOO. Further laboratory studies of this species are needed. The rate-limiting step in the important odd-oxygen destruction cycles is illustrated in Figure 2-14, which shows the loss rate of the odd-oxygen as a function of altitude.

PHOTOCHEMISTRY-TEMPERATURE COUPLING

Since a number of important reactions that significantly influence ozone photochemistry are temperature-dependent, any alteration of the temperature field will affect the photochemistry, and thus the abundance, of ozone. Ozone is, of course, the principal agent for heating the stratosphere by absorption of solar radiation in the near-ultraviolet and visible spectral ranges. In general, a lowering of the temperature slows the reactions that destroy ozone and thus increases its abundance. Lindzen and Will (1973) and Lacis and Hansen (1974) have derived reasonably precise heating rates based upon appropriate solutions to the radiative

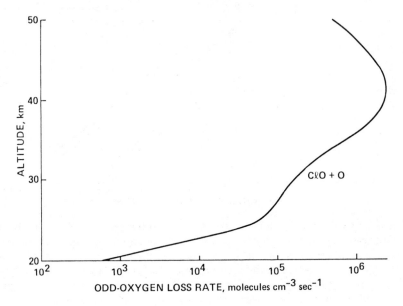

Figure 2-14. Computed odd-oxygen loss rate due to reaction (60).

transfer equation. They have also provided analytic parameterizations of ozone heating based upon solutions of the approximate radiative transfer equation. For example, Lacis and Hansen express the solar flux $A_{l,\,O_3}$, absorbed in the lth layer as

$$A_{l,\,O_3} = \mu_0\{A_{O_3}(x_{l+1}) - A_{O_3}(x_l) + \overline{R}(\mu_0)$$
$$\times\,[A_{O_3}(x_l^*) - A_{O_3}(x_{l+1}^*)]\}, \qquad (2\text{-}74)$$

where x_l is the ozone amount traversed by the direct solar beam, x_l^* is the ozone path traversed by the diffuse radiation illuminating the lth layer from below, μ_0 is the cosine of the angle of incidence of the direct radiation, and $\overline{R}(\mu_0)$ is the effective albedo of the lower atmosphere. The interested reader is referred to the original work for a complete explanation of the radiative transfer equation and expressions for the required parameters.

Cooling of the stratosphere occurs mainly through infrared radiation to space in the 9.6 μm and 15 μm vibrational-rotational bands of carbon dioxide. The first solution of the radiative transfer equation describing the heat energy loss due to CO_2 infrared radiation was obtained by Ramanathan (1974, 1976) who used an appropriate band model. In order to assess the magnitude of the temperature-photochemistry feedback, Luther, Wuebbles, and Chang (1977) have carried out a reasonably complete one-dimensional photochemical model calculation in which they included absorption of solar radiation by ozone (method of Lindzen and Will, 1973), water vapor, and carbon dioxide, as well as cooling by infrared radiation from the same constituents. They also included scattering of solar radiation in the troposphere as well as surface albedo. The mixing ratios of carbon dioxide and water vapor were then varied in order to determine the variation in stratospheric temperature and ozone concentration at various altitudes. However, the water vapor mixing ratio was not allowed to vary with altitude, as would be the case in the real atmosphere as a result of methane oxidation. They found that for a doubling of carbon dioxide abundance, the simulated ozone column density increased by about 3%, with a maximum local change in ozone density of nearly 30% at about 40 km altitude. They also varied the simulated abundances of odd-nitrogen and odd-chlorine (from halocarbons) and obtained the corresponding alterations in temperature and ozone for each simulation. The interested reader is referred to their paper for further details; however, it should be kept in mind that Luther and co-workers used somewhat different chemical kinetic data than are generally accepted at present.

When the atmosphere cools as a result of decreased heating or increased cooling, it contracts. This adjustment to hydrostatic equilibrium alters local rates of formation of ozone by three-body association of O with O_2. The net result (Chandra, Butler, and Stolarski, 1978) is that the effect of temperature feedback is reduced at altitudes above 30 km.

SUMMARY

In this chapter we have discussed the photochemical properties of ozone that govern its production from the photodissociation of molecular oxygen and its destruction by various stratospheric processes. Although the ozone theory of Chapman is still accepted as broadly correct, it has had to be modified in its particulars, especially by the introduction of mechanisms that speed the loss reaction

$$O + O_3 \rightarrow 2\ O_2. \tag{2-12}$$

These new processes, all of which have been suggested over the last twenty years, are catalytic in nature. That is, the reagents other than odd-oxygen that participate in the new scheme are not themselves destroyed by reacting with odd-oxygen but participate over and over before they are lost from the stratosphere. Destruction by reaction with odd-hydrogen (HO_x) was suggested in the mid-1960s, but it could not by itself account for the required ozone loss rate. Crutzen's (1970) suggestion that odd-nitrogen (NO_x) could meet the requirement has come to be generally accepted as the dominant mechanism for ozone catalysis, particularly since the abundance of its stratospheric source, N_2O, is large enough to supply the necessary amount of NO_x. More recently, odd-chlorine (Cl_x) has been proposed as a potentially serious cause of the depletion of stratospheric ozone. However, it is important to note that none of the catalytic cycles that destroy ozone in the laboratory have been *conclusively* demonstrated to occur in the atmosphere. It is probable that the effects of the NO_x cycle have been observed (i.e., the correlation of solar protons with ozone reduction reported by Heath, Krueger, and Crutzen, 1977 and by Reagan et al., 1981 and discussed earlier in this chapter), but more research is needed, especially with respect to the chlorine catalytic cycle. We do not imply that the occurrence of these cycles in the atmosphere is in serious doubt, but they should be unequivocally demonstrated. Chapter 4 deals with the perturbation of stratospheric ozone by both natural and anthropogenic influences in considerable detail.

While chemistry provides the source and ultimate sinks of stratospheric ozone, atmospheric transport processes strongly influence both its distribution and its total abundance. For example, downward transport of ozone from the region where it is formed will deposit it in regions where it is effectively preserved, because loss mechanisms that occur there are very slow. The following chapter treats the details of ozone transport.

REFERENCES

Ackerman, M., D. Frimout, A. Girard, M. Gottignier, and C. Muller, 1976, Stratospheric HCl from infrared spectra, *Geophys. Res. Lett.* **3**:81-83.

Adler-Golden, S. M., E. L. Schweitzer, and J. L. Steinfeld, 1982, Ultraviolet continuum spectroscopy of vibrationally excited ozone, *J. Chem. Phys.* **76**:2201-2209.

Amimoto, S. T., A. P. Force, R. G. Gulotty, Jr., and J. R. Wiesenfeld, 1979, Collisional deactivation of $O(2^1D_2)$ by atmospheric gases, *J. Chem. Phys.* **71**:3640-3647.

Arnold, I., F. J. Comes, and G. K. Moortgat, 1977, Laser flash photolysis: quantum yield of $O(^1D)$ formation from ozone, *Chem. Phys.* **24**:211-217.

Barnes, I., V. Bastian, K. H. Becker, E. H. Fink, and F. Zabel, 1981, Rate constant of the reaction of OH with HO_2NO_2, *Chem. Phys. Lett.* **83**:459-464.

Bates, D. R., and M. Nicolet, 1950, The photochemistry of atmospheric water vapor, *J. Geophys. Res.* **55**:301-327.

Bauer, E., and F. R. Gilmore, 1975, Effect of atmospheric nuclear explosions on total ozone, *Revs. Geophys. and Space Phys.* **13**:451-458.

Borucki, W. J., R. C. Whitten, H. T. Woodward, L. A. Capone, C. A. Riegel, and S. Gaines, 1980, Stratospheric ozone decrease due to chlorofluoromethane photolysis: predictions of latitude dependence, *J. Atmos. Sci.* **37**:686-697.

Brand, J. C. D., K. J. Cross, Jr., and A. R. Hoy, 1978, The Huggins bands of ozone, *Canad. J. Phys.* **56**:327-333.

Brock, J. C., and R. T. Watson, 1980, Laser flash photolysis of ozone: $O(^1D)$ quantum yields in the fall-off region 297-325 nm, *Chem. Phys.* **46**:477-484.

Buijs, H. L., G. L. Vail, G. Tremblay, and D. J. W. Kendall, 1980, Simultaneous measurements of the volume mixing ratio of HCl and HF in the stratosphere, *Geophys. Res. Lett.* **7**:205-208.

Carney, G. D., L. A. Curtiss, and S. R. Langhoff, 1976, Improved potential functions for bent AB_2 molecules: water and ozone, *J. Molec. Spectros.* **61**:371-381.

Castellano, E., and H. J. Schumacher, 1969, Die Kinetik und der Mechanismus des photochemischen Ozonzerfalles im Licht der Wellenlänge 313 μm, *Z. Phys. Chem. (Leipzig)* **65**:62-85.

Chameides, W. L., D. H. S. Stedman, R. R. Dickerson, D. W. Rush, and R. J. Cicerone, 1977, NO_x production in lightning, *J. Atmos. Sci.* **34**:143-149.

Chamberlain, J. W., 1978, *Theory of Planetary Atmospheres: An Introduction to Their Physics and Chemistry*, Academic Press, New York.

Chandra, S., D. M. Butler, and R. S. Stolarski, 1978, Effect of temperature coupling on ozone depletion prediction, *Geophys. Res. Lett.* **5**:199-202.

Chang, J. S., W. H. Duewer, and D. J. Wuebbles, 1979, Atmospheric nuclear tests of the 1950's and 1960's: a possible test of ozone depletion theories, *J. Geophys. Res.* **84**:1755-1765.

Chapman, S., 1930, A theory of upper atmospheric ozone, *R. Meteorol. Soc. Mem.* **3**:103-125.

Chappuis, M. J., 1880, Sur le spectre d'absorption de l'ozone, *Acad. Sci. (Paris) C.R.* **91**:985-986.

Cicerone, R. J., 1981, Halogens in the atmosphere, *Revs. Geophys. Space Phys.* **19**:123-139.

Cox, R. A., and R. G. Derwent, 1976, The ultraviolet absorption spectrum of gaseous nitrous acid. *J. Photochem.* **6**:23-34.

Crutzen, P. J., 1970, The influence of nitrogen oxides on the atmospheric ozone content, *R. Meteorol. Soc. Q. J.* **96**:320-325.

Crutzen, P. J., and C. J. Howard, 1978, The effect of the HO_2 + NO reaction rate constant on one-dimensional model calculations of stratospheric ozone perturbations, *Pure Appl. Geophys.* **116**:497-510.

Crutzen, P. J., I. S. A. Isaksen, and G. C. Reid, 1975, Solar proton events: Stratospheric sources of nitric oxide, *Science* **189**:457-459.

Danti, A., and R. C. Lord, 1959, Pure rotational absorption of ozone and sulfur dioxide from 100 to 200 microns, *J. Chem. Phys.* **30**:1310-1313.

Davenport, J. E., 1980, *Parameters of Ozone photolysis as Functions of Temperature at 280-300 nm,* Rep. FAA-EE-80-44R, SRI International, Menlo Park, Calif. (Also available from National Technical Information Service, Springfield Va.)

DeMore, W. B., R. T. Watson, D. M. Golden, R. F. Hampson, M. Kurylo, C. J. Howard, M. J. Molina, and A. R. Ravishankara, 1982, *Chemical Kinetics and Photochemical Data for use in Stratospheric Modeling, Evaluation Number 5,* Jet Propulsion Laboratory Publication 82-57, Pasadena, Calif.

Duewer, W. H., D. J. Wuebbles, and J. S. Chang, 1977, Effect of NO photolysis on NO_y mixing ratios, *Nature* **265**:523-525.

Fairchild, C. E., E. J. Stone, and G. M. Lawrence, 1978, Photofragment spectroscopy of ozone in the UV region 270-310 nm and at 600 nm, *J. Chem. Phys.* **69**:3632-3638.

Farmer, C. B., O. F. Raper, and R. H. Norton, 1976, Spectroscopic detection and vertical distribution of HCl in the troposphere and stratosphere, *Geophys. Res. Lett.* **3**:13-16.

Frederick, J. E., and R. D. Hudson, 1979, Predissociation of nitric oxide in the mesosphere and stratosphere, *J. Atmos. Sci.* **36**:737-745.

Frederick, J. E., and J. E. Mentall, 1982, Solar irradiance in the stratosphere: Implications for the Herzberg continuum absorption of O_2, *Geophys. Res. Lett.* **9**:461-464.

Goody, R. M., 1964, *Atmospheric Radiation,* Clarendon Press, Oxford.

Graham, R. A., and H. S. Johnston, 1978, The photochemistry of NO_3 and the kinetics of the N_2O_5-O_3 system, *J. Phys. Chem.* **82**:254-268.

Hahn, J., 1979, The cycle of atmospheric nitrous oxide, *R. Soc. (London) Philos. Trans.* **A290**:495-504.

Hahn, J., and C. Junge, 1977, Atmospheric nitrous oxide: A critical review, *Z. Naturforsch.* **32a**:190-214.

Hampson, J., 1964, *Photochemical Behavior of the Ozone Layer,* Canadian Armament Research and Development Establishment TN 1627/64, Canadian Armament Research and Development Establishment, Valcartier, Canada.

Hartley, W. N., 1881, On the absorption spectrum of ozone, *Chem. Soc. (London) J.* **39**:57-60.

Hasson, V., and R. W. Nicholls, 1971, Absolute spectral absorption measurements on molecular oxygen from 2640-1920 Å. II. Continuum measurements 2430-1920 Å, *J. Phys. B: Atom. Molec. Phys.* **4**:1789-1797.

Hay, P. J., and T. H. Dunning, Jr., 1977, Geometries and energies of the excited states of O_3 from *ab initio* potential energy surfaces, *J. Chem. Phys.* **67**:2290-2303.

Hay, P. J., T. H. Dunning, Jr., and W. A. Goddard, 1975, Configuration interaction studies of O_3 and O_3^+. Ground and excited states, *J. Chem. Phys.* **62**:3912-3924.

Hay, P. J., R. T. Pack, R. B. Walker, and E. J. Heller, 1982, Photodissociation of ozone in the Hartley band. Exploratory potential energy surfaces and molecular dynamics, *J. Phys. Chem.* **86**:862-865.

Hayes, E. F., and A. K. Q. Siu, 1971, Electronic structure of the open forms of three membered rings, *Am. Chem. Soc. J.* **93**:2090-2091.

Heath, D. F., A. J. Krueger, and P. J. Crutzen, 1977, Solar proton event: influence on stratospheric ozone, *Science* **197**:886-889.

Henderson-Sellers, A., 1977, Nitric oxide formation by meteoroids in the upper atmosphere, *Atmos. Env.* **11**:864.

Herman, J. R., and J. E. Mentall, 1982, O_2 absorption cross sections from stratospheric solar flux measurements, *J. Geophys. Res.* **87**:8967-8975.

Herzberg, G., 1944, *Molecular Spectra and Molecular Structure, III. Electronic Structure of Polyatomic Molecules*, Van Nostrand, New York.

Howard, C. J., and K. M. Evenson, 1977, Kinetics of the reaction of HO_2 radicals with NO, *Geophys. Res. Lett.* **4**:437-440.

Hudson, R. D., and L. J. Kieffer, 1974, *Absorption Cross Sections of Stratospheric Molecules, The Natural Stratosphere of 1974*, CIAP Monograph No. 1, U.S. Department of Transportation, pp. 5-156 to 5-194; available from National Technical Information Service, Springfield, Va.

Hudson, R. D., and S. H. Mahle, 1972, Photodissociation rates of molecular oxygen in the mesosphere and lower thermosphere, *J. Geophys. Res.* **77**:2902-2914.

Hudson, R. D., V. L. Carter, and E. L. Breig, 1969, Predissociation in the Schumann-Runge bands of O_2: Laboratory measurements and atmospheric effects, *J. Geophys. Res.* **74**:4079-4086.

Huggins, M. L., 1890, On a new group of lines in the photographic spectrum of Sirius, *R. Soc. (London) Proc.* ser. A, **48**:216-217.

Hughes, R. H., 1956, Structure of ozone from the microwave spectrum between 9000 and 45000 Mc, *J. Chem. Phys.* **24**:131-138.

Humphrey, G. L., and R. M. Badger, 1947, The absorption spectrum of ozone in the visible. I. Examination for fine structure, II. Effects of temperature *J. Chem. Phys.* **15**:794-798.

Hunt, B. G., 1966, Photochemistry of ozone in a moist atmosphere, *J. Geophys. Res.* **71**:1385-1398.

Inn, E. C. Y., and Y. Tanaka, 1953, Absorption coefficient of ozone in the ultraviolet and the visible regions, *Opt. Soc. Am. J.* **43**:870-873.

Jesson, J. P., 1980, Release of industrial halocarbons and tropospheric budget, in *Proceedings of the NATO Advanced Study Institute on Atmospheric Ozone: Its Variation and Human Influences*, A. C. Aikin, ed., Rep. No. FAA-EE-80-20,

U.S. Department of Transportation, Federal Aviation Administration, pp. 373-396.

Johnston, H. S., and R. A. Graham, 1974, Photochemistry of NO_x and HNO_x compounds, *Canad. J. Chem.* **52**:1415-1423.

Johnston, H. S., G. Whitten, and J. Birks, 1973, Effect of nuclear explosions on stratospheric ozone, *J. Geophys. Res.* **78**:6107-6135.

Jones, I. T. N., and R. P. Wayne, 1970, The photolysis of ozone by ultraviolet radiation. IV. Effect of photolysis wavelength on primary step, *R. Soc. (London) Proc.* ser. A, **319**:273-287.

Kajimoto, O., and R. J. Cvetanovic, 1976, Temperature dependence of $O(^1D_2)$ production in the photolysis of ozone at 313 nm, *Chem. Phys. Lett.* **37**:533-536.

Kaplan, L. D., M. V. Migeotte, and L. Nevin, 1956, 9.6-micron band of telluric ozone and its rotational analysis, *J. Chem. Phys.* **24**:1183-1186.

Katayama, D. H., 1979, New vibrational quantum number assignment for UV absorption bands of ozone based on the isotope effect, *J. Chem. Phys.* **71**:815-820.

Kuis, S., R. Simonaitis, and J. Heicklen, 1975, Temperature dependence of the photolysis of ozone at 3130 Å, *J. Geophys. Res.* **80**:1328-1331.

Lacis, A. A., and J. E. Hansen, 1974, A parameterization for the absorption of solar radiation in the Earth's atmosphere, *J. Atmos. Sci.* **31**:118-133.

Levine, J. S., R. E. Hughes, W. L. Chameides, and W. E. Howell, 1979, N_2O and CO production by electric discharge: Atmospheric implications, *Geophys. Res. Lett.* **6**:557-559.

Lin, C. L., and W. B. DeMore, 1973/1974, $O(^1D)$ production in ozone photolysis near 3100 Å, *J. Photochem.* **2**:161-164.

Lindzen, R. S., and D. I. Will, 1973, An analytic formula for heating due to ozone absorption, *J. Atmos. Sci.* **30**:513-515.

Liu, S. C., R. J. Cicerone, T. M. Donahue, and W. L. Chameides, 1977, Sources and sinks of atmospheric N_2O and the possible ozone reduction due to industrial fixed nitrogen fertilizers, *Tellus* **29**:251-263.

Logan, J. A., M. J. Prather, S. C. Wofsy, and M. B. McElroy, 1978, Atmospheric chemistry: response to human influence, *R. Soc. (London) Philos. Trans.*, ser. A, **290**:187-234.

Lovelock, J. E., 1975, Natural halocarbons in the air and in the sea, *Nature* **256**:193-194.

Lucchese, R. R., and H. F. Schaeffer III, 1977, Energy separation between the open (C_{2v}) and closed (D_{3h}) forms of ozone, *J. Chem. Phys.* **67**:848-849.

Luther, F. M., and R. J. Gelinas, 1976, Effect of molecular multiple scattering and surface albedo on atmospheric photodissociation, *J. Geophys. Res.* **81**:1125-1132.

Luther, F. M., D. J. Wuebbles, and J. S. Chang, 1977, Temperature feedback in a stratospheric model, *J. Geophys. Res.* **82**:4935-4942.

McElroy, M. B., and J. C. McConnell, 1971, Nitrous oxide: A natural source of stratospheric NO, *J. Atmos. Sci.* **28**:1095-1098.

McElroy, M. B., J. W. Elkins, S. C. Wofsy, and Y. L. Yung, 1976, Sources and sinks for atmospheric N_2O, *Rev. Geophys. and Space Phys.* **14**:143-150.

Magnotta, F., and H. S. Johnston, 1980, Photodissociation quantum yields for the NO_3 free radical, *Geophys. Res. Lett.* **7**:769-772.

Meier, R. R., D. E. Anderson, Jr., and M. Nicolet, 1982, Radiation field in the troposphere and stratosphere from 240-1000 nm. I. General analysis, *Planet. Space Sci.* **30**:923-933.

Molina, L. T., and M. J. Molina, 1979, Chlorine nitrate ultraviolet absorption spectrum at stratospheric temperatures, *J. Photochem.* **11**:139-144.

Molina, L. T., and M. J. Molina, 1980, *Ultraviolet Absorption Cross Sections of HO_2NO_2 Vapor,* Rep. FAA-EE-80-07, University of California, Irvine.

Moortgat, G. K., and E. Kudszus, 1978, Mathematical expression for the O (^{1}D) quantum yields from the O_3 photolysis as a function of temperature (230-320 K) and wavelength (295-320 nm), *Geophys. Res. Lett.* **5**:191-194.

Moortgat, G. K., and P. Warneck, 1975, Relative O (^{1}D) quantum yields in the near UV photolysis of ozone at 298K, *Z. Naturforsch.* **30a**:835-844.

Moortgat, G. R., E. Kudszus, and P. Warneck, 1977, Temperature dependence of O(^{1}D) formation in the near u.v. photolysis of ozone, *Chem. Soc. J. Farraday Trans. II* **73**:1216-1221.

Moortgat, G. K., F. Slemr, W. Seiler, and P. Warneck, 1978, Photolysis of formaldehyde: Relative quantum yields of H_2 and CO in the wavelength range 270-360 nm, *Chem. Phys. Lett.* **54**:444-447.

Murrell, J. N., K. S. Sorbie, and A. J. C. Varandas, 1976, Analytical potential for triatomic molecules from spectroscopic data. II. Application to ozone, *Molec. Phys.* **32**:1359-1372.

Nelson, H. H., W. J. Marinelli, and H. S. Johnston, 1981, The kinetics and product yield of the reaction of OH with HNO_3, *Chem. Phys. Lett.* **78**:495-499.

Nicolet, M., 1971, Aeronomic reactions of hydrogen and ozone, *Mesospheric Models and Related Experiments*, G. Fiocco, ed., Reidel, Dordrecht, Holland.

Nicolet, M., 1975, On the production of nitric oxide by cosmic rays in the mesosphere and stratosphere, *Planet. Space Sci.* **23**:637-649.

Nicolet, M., 1979, Photodissociation of nitric oxide in the mesosphere and stratosphere: Simplified numerical relations for atmospheric photodissociation rates, *Geophys. Res. Lett.* **6**:866-868.

Nicolet, M., and S. Cieslik, 1980, The photodissociation of nitric acid in the mesosphere and stratosphere, *Planet. Space Sci.* **28**:105-115.

Nicolet, M., R. R. Meier, and D. E. Anderson, Jr., 1982, Radiation field in the troposphere and stratosphere. II. Numerical analysis, *Planet. Space Sci.* **30**:935-983.

Oliver, R. C., E. Bauer, H. Hidalgo, K. A. Gardner, and W. Wasylkiwskyj, 1977, *Aircraft Emissions: Potential Effects on Ozone and Climate—A Review and Progress Report. Part I,* FAA-EQ-77-3, DOT-FA 76 WA-3757, U.S. Department of Transportation, Washington, D.C.

Park, C., 1976, Estimates of nitric oxide production for lifting spacecraft reentry, *Atmos. Env.* **10**:309-313.

Park, C., 1978, Nitric oxide production by Tunguska meteor, *Acta Astronautica* **5**:523-542.

Park, C., 1980, Equivalent cone calculation of nitric oxide production rate during space shuttle reentry, *Atmos. Env.* **14**:971-972.

Park, C., and G. P. Menees, 1978, Odd-nitrogen production by meteoroids, *J. Geophys. Res.* **83**:4029-4035.

Philen, D. L., R. T. Watson, and D. D. Davis, 1977, A quantum yield determination of $O(^1D)$ production from ozone via laser flash photolysis, *J. Chem. Phys.* 67:3316-3321.

Prasad, S. S., 1980, Possible existence and chemistry of $ClO·O_2$ in the stratosphere, *Nature* 285:152-154.

Prasad, S. S., 1981, Excited ozone is a possible source of atmospheric N_2O, *Nature* 289:386-388.

Prasad, S. S., and W. M. Adams, 1980, Asymmetrical ClO_3: Its possible formation from ClO and O_2 and its possible reactions, *J. Photochem.* 13:243-252.

Prasad, S. S., and E. C. Zipf, 1982, Atmospheric Sources of Nitrous Oxide—Solar Resonant Excitation of Metastable OH(A) and N_2(A), paper presented at the Fifteenth Informal Conference on Photochemistry, Stanford, Calif.

Ramanathan, V., 1974, A Simplified Stratospheric Radiative Transfer Model: Theoretical Estimates of the Thermal Structure of the Basic and Perturbed Stratosphere, paper presented at the Second International Conference on the Environemntal Impact of Aerospace Operations in the High Atmosphere, American Meteorological Society, American Insitute of Aeronautics and Astronautics, San Diego, Calif.

Ramanathan, V., 1976, Radiative transfer within the earth's troposphere and stratosphere: A simplified radiative-convective model, *J. Atmos. Sci.* 33:1330-1346.

Raper, O. F., C. B. Farmer, R. A. Toth, and B. D. Robbins, 1977, The vertical distribution of HCl in the stratosphere, *Geophys. Res. Lett.* 4:531-534.

Reagan, J. B., R. E. Meyerhott, R. W. Nightingale, R. C. Gunton, R. G. Johnson, J. E. Evans, W. L. Imhof, D. F. Heath, and A. J. Krueger, 1981, Effects of the August 1972 solar particle events on stratospheric ozone, *J. Geophys. Res.* 86:1473-1494.

Reid, G. C., J. R. McAfee, and P. J. Crutzen, 1978, Effects of intense stratospheric ionization events, *Nature* 257:489-492.

Shand, W., Jr., and R. A. Spurr, 1943, The molecular structure of ozone, *Am. Chem. Soc. J.* 65:179-181.

Shardanand and A. D. Prasad Rao, 1977, Collision-induced absorption of O_2 in the Herzberg continuum, *J. Quant. Spectrosc. Radiat. Transfer* 17:433-439.

Shih, S., R. J. Buenker, and S. D. Peyerimhof, 1974, Theoretical investigation of the cyclic conformer of ozone, *Chem. Phys. Lett.* 28:463-470.

Shimazaki, T., and L. C. Helmle, 1977, A simplified method for calculating the atmospheric heating rate by absorption of solar radiation in the stratosphere and mesosphere, *NASA Technical Paper 1398*, NASA-Ames Research Center, Moffett Field, Calif.

Simons, J. W., R. J. Paur, H. A. Webster III, and E. J. Bair, 1973, Ozone ultraviolet photolysis. VI. The ultraviolet spectrum, *J. Chem. Phys.* 59:1203-1208.

Simonaitis, R., S. Braslavsky, J. Heicklen, and M. Nicolet, 1973, Photolysis of O_3 at 3130 Å, *Chem. Phys. Lett.* 19:601-603.

Singh, H. B., L. J. Salas, and H. Shigeishi, 1979, The distribution of nitrous oxide (N_2O) in the global atmosphere and the Pacific Ocean, *Tellus* 31:313-320.

Singh, H. B., L. J. Salas, H. Shigeishi, and E. Scribner, 1979, Atmospheric halocarbons, hydrocarbons and sulfur hexafluoride: global distributions, sources and sinks, *Science* 203:899-903.

Smith, F. L., and C. Smith, 1972, Numerical evaluation of Chapman's grazing incidence integral ch(X,χ), *J. Geophys. Res.* 77:3592-3597.

Sparks, R. K., C. R. Carlson, K. Shobatake, M. L. Kowalczyk, and Y. T. Lee, 1980, Ozone photolysis: A determination of the electronic and vibrational state distributions of primary products, *J. Chem. Phys.* 72:1401-1402.

Stedman, D. H., and R. E. Shetter, 1983, The global budget of atmospheric nitrogen species, in *Trace Constituents; Properties, Transformation and Fates*, S. E. Schwartz, ed., Wiley, New York, 411-454.

Stolarski, R. S., and R. J. Cicerone, 1974, Stratospheric chlorine: A possible sink for ozone. *Canad. J. Chem.* 52:1610-1615.

Swanson, N., and R. J. Celotta, 1975, Observation of excited states in ozone near the dissociation limit, *Phys. Rev. Lett.* 35:783-785.

Thorne, R. M., 1980, The importance of energetic particle precipitation on the chemical composition of the middle atmosphere, *Pure Appl. Geophys.* 118:128-151.

Thuneman, K. H., S. D. Peyerimhof, and R. J. Buenker, 1978, Configuration interaction calculations for the ground and excited states of ozone and its positive ion: Energy locations and transition probabilities, *J. Molec. Spectrosc.* 70:432-448.

Trambarulo, R., S. N. Ghosh, C. A. Burrus, Jr., and W. Gordy, 1953, Molecular structure, dipole moment, and g factor of ozone from its microwave spectrum, *J. Chem. Phys.* 21:851-855.

Trevor, P. L., G. Black, and J. R. Barker, 1982, Reaction rate constant for OH + HOONO$_2$ → products over the temperature range 246-324 K, *J. Phys. Chem.* 86:1661-1669.

Tuck, A. F., 1976, Production of nitrogen oxides by lightning discharges, *R. Meteorol. Soc. Q. J.* 102:749-755.

Turco, R. P., 1975, Photodissociation rates in the atmosphere below 100 km, *Geophys. Surv.* 2:153-192.

Turco, R. P., and R. C. Whitten, 1977, The NASA-Ames Research Center one- and two-dimensional stratospheric models. I. The one-dimensional model, *NASA Technical Paper 1002*, NASA-Ames Research Center, Moffett Field, Calif.

Turco, R. P., R. C. Whitten, I. G. Poppoff, and L. A. Capone, 1978, SST's, nitrogen fertilizer and stratospheric ozone, *Nature* 276:805-807.

Vigroux, E., 1953, Contribution a l'étude experimentale de l'absorption de l'ozone, *Ann. de Phys.* 8:709-762.

Von Rosenberg, C. W., Jr., and D. W. Trainor, 1975, Excitation of ozone formed by recombination. II., *J. Chem. Phys.* 63:5348-5353.

Watson, A. J., J. E. Lovelock, and D. H. Stedman, 1980, The problem of atmospheric methyl chloride, in *Proceedings of the NATO Advanced Study Institute on Atmospheric Ozone: Its Variation and Human Influence*, A. C. Aikin, ed., Rep. No. FAA-EE-80-20, U.S. Department of Transportation, Federal Aviation Administration, 365-372.

Weinstock, E. M., M. J. Phillips, and J. G. Anderson, 1981, *In situ* observations of ClO in the stratosphere: A review of recent results, *J. Geophys. Res.* 86:7273-7278; (See also Anderson, J. G., H. J. Grassl, R. E. Shetter, and J. J.

Margitan, 1980, Stratospheric free chlorine measured by balloon-borne *in situ* resonance fluorescence, *J. Geophys. Res.* 85:2869-2887.)

Weiss, R. F., 1981, The temporal and spatial distribution of tropospheric nitrous oxide, *J. Geophys. Res.* 86:7185-7195.

Whitten, R. C., W. J. Borucki, and R. P. Turco, 1975, Possible ozone depletions following nuclear explosions, *Nature* 257:38-39.

Whitten, R. C., W. J. Borucki, V. R. Watson, T. Shimazaki, H. T. Woodward, C. A. Riegel, L. A. Capone, and T. Becker, 1977, The NASA-Ames Research Center one- and two-dimensional stratospheric models. II. The two-dimensional model, *NASA Technical Paper 1003*, NASA-Ames Research Center, Moffett Field, Calif.

Whitten, R. C., W. J. Borucki, H. T. Woodward, L. A. Capone, C. A. Riegel, R. P. Turco, I. G. Poppoff, and K. Santhanam, 1981, Implications of smaller concentrations of stratospheric OH: A two-dimensional model study of ozone perturbations, *Atmos. Env.* 15:1583-1598. (See also Whitten, R. C., W. J. Borucki, H. T. Woodward, L. A. Capone, and C. A. Riegel, 1983, Revised predictions of the effect on stratospheric ozone of increasing atmospheric N_2O and chlorofluoromethanes: A two-dimensional model study, *Atmos. Env.* 17:1995-2000.)

Whitten, R. C., W. J. Borucki, C. Park, L. Pfister, R. P. Turco, L. A. Capone, C. A. Riegel, and T. Kropp, 1982, The satellite power system: Assessment of the environmental impact on middle atmosphere composition and on climate, *Space Solar Power Rev.* 3:195-221.

Williams, W. J., J. J. Kosters, A. Goldman, and D. G. Murcray, 1976, Measurements of the stratospheric mixing ratio of HCl using an infrared absorption technique, *Geophys. Res. Lett.* 3:383-385.

Wilson, C. W., Jr., and D. G. Hopper, 1981, Theoretical studies of the ozone molecules. I. Ab initio MCSCF/CI potential energy surfaces for the X^1A_1 and a $^3B_2$ states, *J. Chem. Phys.* 74:595-607.

Wine, P. H., A. R. Ravishankara, N. M. Kreutter, R. C. Shah, J. M. Nicovich, R. L. Thompson, and D. J. Wuebbles, 1981, Rate of reaction of OH with HNO_3, *J. Geophys. Res.* 86:1105-1112.

Wongdontri-Stuper, W., R. K. M. Jayanti, R. Simonaitis, and J. Heicklen, 1979, The Cl_2 photosensitized decomposition of O_3: The reactions of ClO and OClO with O_3, *J. Photochem.* 10:163-186.

World Meteorological Organization, 1981, *The Stratosphere 1981. Theory and Measurements*, WMO Global Ozone Research and Monitoring Project, Rep. No. 11, Geneva, Switzerland.

Wright, J. S., 1973, Theoretical evidence for a stable form of cyclic ozone and its chemical consequences, *Canad. J. Chem.* 51:139-146.

Wright, J. S., S. Shih, and R. J. Buenker, 1980, *Ab initio* potential energy surface for ozone decomposition, *Chem. Phys. Lett.* 75:513-518.

Wulf, O. R., 1930, The band spectrum of ozone in the visible and photographic infrared, *Natl. Acad. Sci. (USA) Proc.* 16:507.

Yung, Y. L., 1976, A numerical method for calculating the mean intensity in an inhomogeneous Rayleigh scattering atmosphere, *J. Quant. Spectrosc. Radiat. Transfer* 16:755-761.

Zafonte, L. P., P. L. Rieger, and J. R. Holmes, 1975, Some aspects of the atmospheric chemistry of nitrous acid, Pacific Conference on Chemistry and Spectroscopy, Asilomar, Calif.

Zander, R., 1981, Recent observations of HF and HCl in the upper atmosphere, *Geophys. Res. Lett.* **8**:413-416.

Zellner, R. E., and V. Handwerk, 1982, Kinetics of ClO with NO_2 and O_2, paper presented at International Symposium of Chemical Kinetics Related to Atmospheric Chemistry, Tsukuba Center for Institutes, Tsukuba, Ibaraki, Japan.

Zittel, P. F., and D. D. Little, 1980, Photodissociation of vibrationally excited ozone, *J. Chem. Phys.* **72**:5900-5905.

Chapter 3

Ozone Transport

Edwin F. Danielsen
Space Science Division
NASA-Ames Research Center

In principle, atmospheric transport is conceptually simple. Only two processes are involved: advection by organized mean motions and dispersion by disorganized, random deviations from the mean motions. When a finite amount of a trace species i is injected or produced in the atmosphere, *advection* describes the movement of the mass M_i including the movement of its center of mass, its deformation or change in shape relative to its center of mass, and its change in volume due to changes in pressure; but there is no diffusive spreading into its environment. Therefore, since $M_i = M_a \overline{\chi}_i$, where M_a is the mass of air into which M_i is mixed and $\overline{\chi}_i$ is the average mixing ratio, advection conserves all three quantities and is a reversible transport process. Conversely, *dispersion* describes the diffusive spreading into the surrounding environment that increases M_a and, since M_i is conserved, decreases $\overline{\chi}_i$. It is an irreversible transport process. If there is no removal process, $\overline{\chi}_i$ will decrease to an absolute limit when M_a equals the mass of the atmosphere.

The smallest scale organized motions are those to which the equations of fluid dynamics apply; that is, they represent mean velocities for a cubic millimeter or cubic centimeter, the larger of which will be referred to here as an elemental volume. The disorganized, random deviations are then due to the molecular velocities. Since these are always present at temperatures warmer than absolute zero, dispersion, in this case molecular dispersion, is always present and always actively reducing $\overline{\chi}_i$ for a fixed amount of M_i. The rate of reduction is small, but it becomes efficient when the fluid is stirred.

In the atmosphere, stirring is generated by many mechanisms, including buoyantly driven verticle circulations and wind shears produced by a broad spectrum of internally propagating waves. The wave motions are reversible and, as such, represent advections and displacements by organized mean motions. However, they intermittently and locally trigger one of several instabilities that effectively mix or stir the atmosphere. Therefore, in the atmosphere the distinction blurs between advection and dispersion, reversible and irreversible transport. It becomes a source of confusion and controversy.

The problem is a direct consequence of the low density of atmospheric observations. To be compatible with even the most numerous measurements, those of meteorological variables, we must integrate the equations of fluid dynamics over large bulk volumes to establish representative mean values. For global analyses and descriptions, a typical bulk volume extends over 10^5 km^2 in the horizontal (3 degrees of latitude squared) and 1 km or more in the vertical for a huge lower limit of 10^{20} cm^3 per bulk volume. In other words, each bulk volume contains more elemental volumes than each elemental volume contains molecules. When we consider sparsely measured trace species such as ozone, the integrals must be considerably larger and extend over long time periods. One frequently encounters integrals over all longitudes and over a month or season of duration τ that yield zonal-temporal mean distributions in latitude ϕ and height z. An additional integration over all latitudes or over Northern Hemispheric latitudes and all seasons yields global mean profiles as a function of z suitable for comparisons with one dimensional (1-D) models. In this chapter we will denote a local bulk mean by ($\overline{}$), a zonal-temporal mean by ($\overline{}$)$^{\lambda\tau}$, a global mean by ($\overline{}$)$^{\phi\lambda\tau}$, and an atmospheric mean value by ($\overline{}$)$^{\phi\lambda z\tau}$.

To discuss the transport of ozone and other trace gases or aerosols we shall first compare the appropriate mass or number continuity equation for a bulk system to that for an elemental system. Let $\rho_i = \rho\chi_i$, where ρ denotes the mass or number density of air molecules. Then for a local bulk system,

$$\frac{\partial \overline{\chi}_i}{\partial t} = -\overline{\mathbf{V}} \cdot \nabla \overline{\chi}_i - \frac{\nabla \cdot \rho \overline{\mathbf{V}'\chi_i'}}{\overline{\rho}} + \frac{\gamma \nabla^2 \overline{\rho}\,\overline{\chi}_i}{\overline{\rho}} + \overline{s}_i, \qquad (3\text{-}1)$$

while for an elemental system

$$\frac{\partial \chi_i}{\partial t} = -\mathbf{V} \cdot \nabla \chi_i + \gamma \nabla^2 \chi_i + s_i, \qquad (3\text{-}2)$$

where γ is the molecular diffusion coefficient, treated here as a constant, and s_i is the nondimensional source or sink. The major difference between (3-1) and (3-2), other than the overbar to denote the mean, is the additional deviatory correlation term in which the deviations from the means are denoted by a prime. In (3-2), the first term on the right describes the reversible local change due to advection. The second term describes the irreversible change due to molecular dispersion. Equation (3-1) has similar terms with similar significance, but the extra term, expressing the local change is $\overline{\chi}_i$ due to the divergence of the deviatory flux, although derived directly from the reversible elemental motions, contributes to both reversible and irreversible transports.

Equation (3-1) is expressed in Eulerian or partial derivative form, whereby the bulk grid volumes remain fixed while the fluid moves through them from one volume to the next. The reference system is completely open to all scales of motion including the molecular motions. If, however, we transfer the mean advection term to the left side of (3-1) and combine it with the partial time derivative, we have a form suitable for Lagrangian averaging. The left side is then a total derivative of the mean; in other words, it is the total derivative of the averaging integral.

The averaging integral moves with the velocity $\overline{\mathbf{V}}$, the velocity of the center of mass of the bulk system. If we constrain each bulk system to contain a constant mass we have the bulk analog to the elemental Lagrangian or substantial derivative. It represents a closed system for all scales of motion larger than that which defines $\overline{\mathbf{V}}$ but is open to all smaller scales, just as an elemental Lagrangian system is open to molecular motions.

This condition is made clear if we use Gauss's divergence theorem to convert the first two of the remaining terms on the right side of (3-1) from a volume to a surface integral. Then (3-1) becomes

$$\frac{d\overline{\chi}_i}{dt} = - \oint \frac{\rho \mathbf{V}' \chi_i' \cdot d\mathbf{A}}{M_a} + \gamma \oint \frac{\rho \nabla \chi_i \cdot d\mathbf{A}}{M_a} + \overline{s}_i \qquad (3\text{-}3)$$

where $d\mathbf{A}$ is a differential surface area of the bulk system and M_a is the total mass of air in the system. Equation (3-3) shows explicitly that as we follow its center of mass, the mean mixing ratio $\overline{\chi}_i$ changes because of a net deviatory flux $\mathbf{V}'\chi'_i$ across the surface; a molecular dispersion across the same surface; and a net source or sink inside the system. The deviatory flux greatly exceeds the molecular dispersion; thus, the latter is usually neglected and its effects are incorporated in the expression for the former.

It is important for our description of ozone transport to recognize that the deviatory flux cannot be reduced to zero unless we reduce our averaging integral to an elemental volume or mass. This reduction is practically impossible, because it would require 10^{20} times the current number of observations. However, when the bulk volume is kept as small as possible, we can usually obtain reasonable first approximations to Lagrangian trajectories by integrating momentum equations analogous to (3-3) while neglecting the deviatory fluxes of momentum.

From (3-1) and (3-3) we conclude that an Eulerian integral is open to all scales of motion and a Lagrangian integral is open to all scales smaller than that of the integral but closed to all larger scales. Both reference systems describe the same transport processes, but from a different perspective. Relative to the fixed Eulerian reference, the flow at any instant is described by streamlines — lines tangent to the velocity vectors of different air parcels at constant time. On the other hand, the moving Lagrangian references generate trajectories — lines tangent to the velocity vectors of the same air parcel at successive times. We shall use both reference systems in this chapter, preferring local bulk Lagrangian systems for physical descriptions in three-dimensional space and Eulerian ones for all others, especially for quantitative descriptions in one- and two-dimensional space.

As we reduce the independent spatial variables from three to two to one the relative importance of the first and second terms on the right-hand side of (3-1) changes progressively. In 3-D, with local bulk integrals, advection dominates; in 2-D, the advective and deviatory terms compete and the latter tend to dominate; while in 1-D, only the deviatory term remains. Due entirely to the lack of sufficient observations, the deviatory terms must be statistically related to gradients of mean quantities.

A major reason for computing trajectories with Lagrangian integrals is to derive representative vertical velocities for large-scale atmospheric motions. These velocities, too small to be directly measured, are of fundamental importance to transport because they are sustained for long time periods and hence produce relatively large vertical displacements. In constructing trajectories we use potential temperature θ rather than height as the vertical coordinate. Since θ is quasi-conservative we can take advantage of the adiabatic approximation to compute net vertical displacements at constant θ. These terms and concepts are discussed below along with another important quasi-conservative quantity called the potential vorticity, denoted here by S. As will be shown, S is positively correlated with the ozone mixing ratio below the stratospheric ozone maximum. This correlation provides a rational basis for extending limited ozone observations to the global scale and permits us to separate effects produced by transport from those produced by photochemical-chemical processes. Needless to say, it is the most important atmospheric variable for trace studies in general and ozone in particular.

HORIZONTAL-VERTICAL TRANSPORT
AND CONSERVATIVE QUANTITIES

The atmosphere is a thin film surrounding a large spheroidal earth. Typical horizontal dimensions exceed vertical by about 1000 to 1. It is not surprising, therefore, to find similar ratios for the horizontal-to-vertical speeds of the large scale motions. For example, horizontal speeds range from 10 to 100 ms^{-1} while their corresponding vertical speeds are only 1 to 10 cm s^{-1}. The latter, in addition to being difficult to measure, are extremely difficult to calculate. Generally, numerical models lead to noisy solutions that have to be smoothed or filtered before they are useful. Resolution degenerates in the process, rendering the vertical velocities suspect for flux computations.

One method which has proven to be computationally stable for large scale transport (both horizontal and vertical) is based on 12-hr trajectories computed from the winds and energies on surfaces of constant θ. Potential temperature θ is derived from simultaneous measurements of temperature T and pressure p, according to

$$\theta = T (p_0/p)^{0.286} \tag{3-4}$$

where T is expressed in kelvins and p in millibars, with $p_0 = 1000$ mb. θ is conserved when no heat is transferred to or from the moving air parcel, a so-called adiabatic process. Since diatomic nitrogen and oxygen, the dominant components of dry air, are poor conductors and radiators, the adiabatic assumption is a reasonable first approximation for trajectory analysis. Diabatic rates are small because they are produced by minority gases, primarily water vapor, carbon dioxide, and ozone.

Under the adiabatic assumption, θ becomes a constant Lagrangian coordinate. Since we can analyze the height of each θ surface every 12 hrs, the net 12-hr displacements along the trajectories yield representative mean vertical velocities. This method automatically filters out the shorter-

Figure 3-1 (on next page). *(A):* Isentropic analysis, $\theta = 305$ K. Stream function and isotachs (kts), 0000 GMT, 30 November 1963. *(B):* Isentropic analysis, $\theta = 305$ K. Stream function and isotachs 1200 GMT, 30 November 1963. *(C):* Isentropic analysis, $\theta = 305$ K. Twelve-hour trajectories. Open circles are mid-time positions where vertical velocities apply. *(D):* Isentropic analysis, $\theta = 305$ K. Twelve-hour mean vertical velocities (cm s^{-1}). Light grey tone indicates cloud cover greater then 50%; dark grey tone indicates precipitation. *(A through C from and D after Danielsen, E. F., R. Bleck, R. Adler, D. Deaven, C. Dey, V. Kousky, and F. Sechrist, 1966, Research in Four-Dimensional Diagnosis of Cyclonic Storm Cloud Systems, AFCRL-66-30, Air Force Cambridge Research Laboratory Rep. No. 2, Hanscom Field, Mass.)*

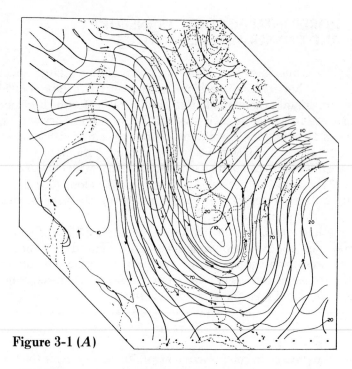

Figure 3-1 (A)

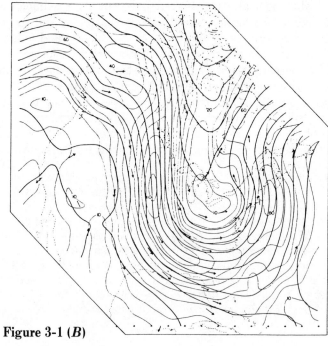

Figure 3-1 (B)

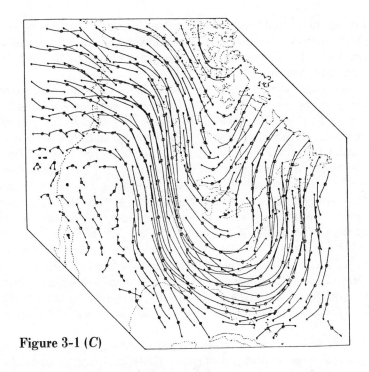

Figure 3-1 (*C*)

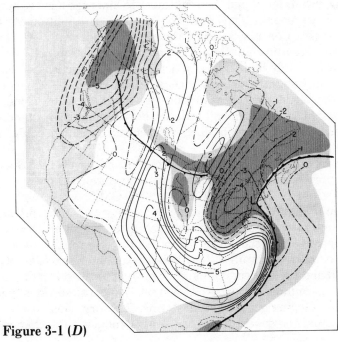

Figure 3-1 (*D*)

period oscillations that tend to make instantaneous velocities noisy. An example of computer-derived trajectories and vertical velocities is presented in Figure 3-1 (*A*–*D*). Computer algorithms were written by R. Bleck using a method developed by Danielsen (1961).

Figures 3-1 (*A*) and (*B*) show successive 12-hr contours of the stream function ψ on the $\theta = 305$ K surface. Gradients of ψ determine both the direction and the speed of the horizontal winds used in the trajectory computations. Both figures also include isotachs, lines of constant wind speed, which indicate a major jet stream, speeds 70–100 kts (35–50 ms^{-1}), which is oriented almost parallel to the flow in a ridge-trough-ridge pattern. West of the trough the jet separates stratospheric air to its north from tropospheric air to its south. East of the trough the separation is much more complex, as we shall soon see from the trajectories. The trajectories in Figure 3-1 (*C*) resemble the pattern of streamlines from which they are derived, but the speed shears generate large differences in the displacements. These displacements increase to a maximum in the jet where the air moves from southern Canada to southern United States in just 12 hrs. As this air moves from the northwest to the southeast (northwest winds) it undergoes strong descent, as indicated in Figure 3-1 (*D*).

As the air descends it maintains its constant θ, but the temperature increases due to adiabatic compression. This warming, in turn, evaporates any initial clouds and then continues to lower the relative humidity. Thus the descending air should be cloudless, a condition verified in Figure 3-1 (*D*). Conversely, where the air is ascending, $w > 0$, there are clouds and even precipitation. The positive correlation between w and χ_{H_2O} is remarkably high, which indicates that the vertical velocity patterns are only slightly inclined in the vertical. Figure 3-1 (*D*), of course, depicts the vertical velocities for only one θ surface; thus, clouds could exist either at lower or higher θ surfaces. For example, the east-west band of clouds and precipitation along the central Canadian border is well below the $\theta = 305$ K surface.

The large area of precipitation over the northeastern United States and southeastern Canada is associated with an intense November storm centered close to New York City at 0600 GMT, 30 November 1963. Twenty-four hours earlier this storm was forming over western Tennessee. The maximums of ascending and descending motions were then east and west respectively of the storm center. As the storm intensified rapidly and moved northeastward to New York the ascending motion rotated to the north and the descending motions rotated to the south of the storm center. This counter-clockwise rotation relative to the storm is typical of rapidly developing extratropical storms. The dry, cloudless air then wraps around the south side of the storm producing the characteristic comma-shaped cloud pattern seen in satellite photographs. The source of the dry air is shown explicitly in Figure 3-2. A set of almost equally spaced points along the boundary between the dry and the moist air were selected as final points, and 36-hr trajectories were computed backward in time to determine their corresponding initial points.

The six northernmost points trace back to the stratosphere on the north side of the major jet over western Canada. Initial pressures range from 295-330 mb, final pressures from 400-550 mb. The next three come from the upper troposphere: initial pressures range from 355-385 mb, final pressures from 715-735 mb. The final six trace back to the middle troposphere with initial pressures 455-620 mb, final pressures from 700-800 mb. In 36 hrs each bulk parcel moves downward to higher pressures. The stratospheric parcels move into the upper and middle troposphere and the upper tropospheric parcels move to the middle and lower troposphere.

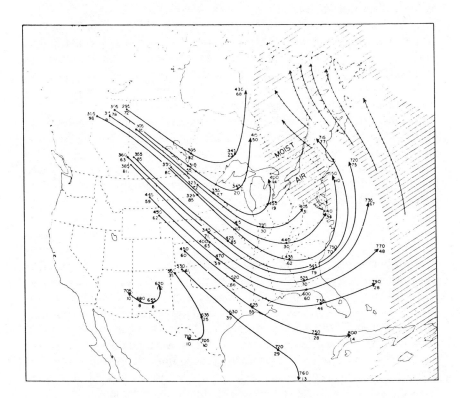

Figure 3-2. Continuous 36-hr trajectories computed backwards in time from positions at 1200 GMT, 30 November 1963. Pressures (mb) and wind speeds (kts) are plotted at successive 12-hr positions. *(From Danielsen, E. F., 1974, Review of trajectory methods, Adv. Geophys. **18**:85; copyright © 1974 by Academic Press, Inc.)*

None of these parcels will return to their initial heights because they are moving faster than the boundary of the moist air. They will overtake and mix with this moist air rendering the displacements irreversible.

Their fate is not so evident in the Eulerian reference frame. The stream functions describe a large-scale wave that would apparently yield reversible transports, as would analyses at constant pressure. It is the long-period trajectories of the Lagrangian reference that depict the large-scale deformations and the inevitable irreversibility. This example is included here to demonstrate the physical insights gained from trajectory analyses. Case studies of this type were used to determine how the tropopause deformed during large-scale cyclogenesis. The process, called tropopause folding, was proposed by Reed and Danielsen (1959). It was predicted and verified by Danielsen (1964, 1968) during Project Springfield, April 1963. Aircraft equipped to capture and measure radioactive aerosols, products of high-yield nuclear bomb tests, were flown through the folded structures to determine whether or not the air in the fold was of stratospheric origin. The results were clear. Indeed, on the basis of the experimental evidence we can predict with certainty that ozone from the lower stratosphere is transported into the troposphere by the process of tropopause folding. With the development of fast-responding ozone sensors for aircraft during the 1970s the ozone transport was confirmed.

Another result of the aircraft experiments was verification of the assumption that the stratospheric air could be identified in the troposphere by its large values of potential vorticity.

Potential vorticity, introduced by Ertel (1942) as a quasi-conservative scalar, can be reduced to its simplest form when θ is used as a vertical coordinate. For each elemental mass,

$$S = a \frac{\partial \theta}{\partial z} \eta_\theta, \tag{3-6}$$

where a is the specific volume, $\partial\theta/\partial z$ is a measure of the static stability (the term used conventionally to distinguish stratosphere from troposphere), and η_θ is the absolute vorticity (a measure of inertial stability). Static stability applies to an air parcel displaced vertically. When its environment is stable it will be forced back to its initial position. When it is unstable it will be accelerated away and lead to vertical mixing. Similarly, inertial stability applies to a parcel displaced horizontally at constant θ. When its environment is inertially unstable, horizontal mixing is the most probable result of even an infinitesimal displacement.

Vorticity is a vector quantity, but in (3-6) only its vertical component is relevant. This component, in turn, includes the vorticity due to motions relative to the earth — the observed winds — and the vorticity due to the earth's rotation. Thus,

$$\eta_\theta = \left(\frac{V}{R} - \frac{\partial V}{\partial n}\right)_\theta + f, \qquad (3\text{-}7)$$

where V is the speed of the wind, R is the radius of curvature of the streamlines ψ, $-\partial V/\partial n$ is the speed shear normal to the streamlines (positive n is to the left facing downwind), and $f = 2\omega \sin \phi$, with ω the angular speed of rotation about the polar axis and ϕ the latitude angle. The θ subscript denotes that the relative vorticity is computed from map analyses at constant θ. The schematics in Figure 3-3 illustrate the basic conditions encountered in atmospheric flows.

The streamlines ψ in Figures 3-1 (A) and (B) are merely combinations of these four examples. To the left or poleward of a jet axis, η_θ is large because $\partial V/\partial n$ is positive and augments f. To the right, η_θ is small or negative because the shear vorticity is negative and subtracts from f.

It follows that on the $\theta = 305$ K surface illustrated in Figure 3-1 inertial stability is much larger poleward than equatorward of the jet. Also, the static stability is larger poleward than equatorward of the jet, so S is very much larger in the stratosphere than it is in the troposphere. Since S is conserved for adiabatic motions, a tropopause defined by a critical value of S will simply fold during large scale cyclogenesis as shown in Figure 3-4 (b).

When the tropopause folds, it becomes vertical at the jet axis. Since $\partial\theta/\partial z$ is then continuous across the jet, the conventional definition of the

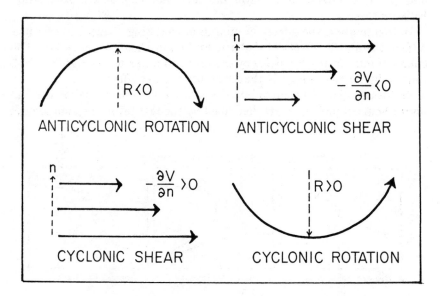

Figure 3-3. Schematic examples of relative vorticity.

tropopause fails, and some authors refer to a broken tropopause (Fig. 3-4 c). This latter concept is potentially misleading, because the horizontal vorticity gradient is as effective a barrier to exchange as is a vertical gradient of $\partial\theta/\partial z$. Also, the transport from stratosphere to troposphere occurs not across the jet axis but beneath it. Similarly, if tropospheric air enters the stratosphere, the transport is not across the jet axis but above it. Furthermore, the transports are not a diffusive type of leaking, as implied by the break, but an organized mass transfer with well-defined boundaries, as shown in Figure 3-4 (b).

Air within the folded tropopause layer, toned grey in Figure 3-4 (b), has large mixing ratios of stratospheric radioactive isotopes, stratospheric trace gases, such as ozone, and large values of potential vorticity, that is, $\overline{S}$ and $\overline{\chi}_{O_3}$ are positively correlated as are their gradients $\nabla\overline{S}$ and $\nabla\overline{\chi}_{O_3}$. Although S and χ_{O_3} both have high altitude, stratospheric sources, and low altitude and/or surface sinks, the sources differ latitudinally: S is predominantly generated at high latitudes, χ_{O_3} at low latitudes. As the ozone and potential vorticity leave their source regions and are transported downward toward the lower stratosphere, S and χ_{O_3} are quasi-conserved, but change following an elemental parcel due to horizontal and vertical mixing. The mixing, which is adiabatic and irreversible, produces the positive correlation and subsequently maintains it as the folded layer entrains tropospheric air along its boundaries. The entrained air is deficient in ozone and potential vorticity; thus it decreases both $\overline{S}$ and $\overline{\chi}_{O_3}$ as it mixes in the layer but does not change the correlation between them.

In this respect, the effects of mixing in changing $\overline{S}$ and $\overline{\chi}_{O_3}$ are quite different from the effects of diabatic and photochemical processes. The effects of mixing are described by the first two terms on the right of (3-3), that is, by surface integrals that evaluate the net effect of air parcels or molecules moving out from or into the volume. The effects of diabatic and photochemical processes are described by the last term, the source-sink

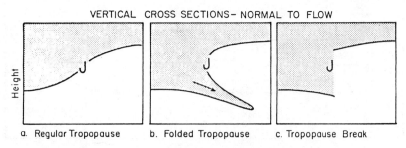

VERTICAL CROSS SECTIONS− NORMAL TO FLOW

Height

a. Regular Tropopause   b. Folded Tropopause   c. Tropopause Break

Figure 3-4. Schematic examples of continuous and broken tropopause.

term, which is evaluated by integrating over every elemental volume. A gradient of diabatic heating can change S and $\overline{S}$ without changing $\overline{\chi}_{O_3}$. Conversely, photochemical processes can change χ_{O_3} and $\overline{\chi}_{O_3}$ without changing $\overline{S}$. Therefore, these nonconservative processes can change the correlation between $\overline{S}$ and $\overline{\chi}_{O_3}$. In general, if the correlation does change during transport in the lower stratosphere or troposphere, nonconservative processes are probably responsible.

ONE-DIMENSIONAL TRANSPORT, VERTICAL FLUX

Ozone transport is reduced to a problem in only one dimension by integrating over all latitudes, longitudes, and seasons. The sole remaining independent variable is height. If the $\overline{\chi}_{O_3}^{\phi\lambda\tau}$ profile has a well-defined maximum or minimum, then a net vertical flux or vertical transport is implied, directed from a maximum to an adjacent minimum. Since the integrations eliminate all mean motions from consideration, the entire vertical transport must be due to correlations between vertical velocity deviations w' and ozone mixing ratio deviations χ'_{O_3}, that is, (3-1) reduces to

$$\frac{\partial \overline{\chi}_{O_3}^{\phi\lambda\tau}}{\partial t} = -\frac{1}{\overline{\rho}} \frac{\partial (\overline{\rho}\, \overline{w'\chi'}_{O_3}^{\phi\lambda\tau})}{\partial z} + \overline{s}_{O_3}^{\phi\lambda\tau}. \tag{3-8}$$

From the previous discussions it should be obvious that we lack the observational capability to evaluate the deviatory flux and its vertical divergence. However, we can estimate the flux of ozone from the stratosphere to the troposphere using observed correlations with the radioactive aerosols, specifically strontium-90, and the deposition of ^{90}Sr measured by the nuclear fallout network. On the basis of these data a net downward flux of 7.8×10^{10} ozone molecules cm^{-2} s^{-1} is expected for the northern hemisphere (Danielsen and Mohnen, 1977). This estimate represents an annual average to be consistent with the integration over all seasons.

By analogy to molecular diffusion, we make the reasonable assumption that the net flux is directed down gradient and set

$$\overline{F}_z^{\phi\lambda\tau}(O_3) = \overline{\rho}\, \overline{w'\chi'}_{O_3}^{\phi\lambda\tau} = -\overline{\rho} K_z \frac{\partial \overline{\chi}_{O_3}^{\phi\lambda\tau}}{\partial z}, \tag{3-9}$$

where K_z is a proportionality factor called the vertical diffusion coefficient. The problem for the transport specialist is to determine a representative vertical profile of K_z, if, indeed, one does exist. It is highly improbable that there is a unique K_z profile applicable to all trace species, but the assumption may apply within a reasonable range of uncertainty. Under these circumstances we could justify the use of (3-9) in 1-D numerical photochemical models.

The author attempted to derive a representative K_z profile and to establish the uncertainty limits while he was a member of the National Academy of Sciences panel on Stratospheric Chemistry and Transport. Three K_z profiles were derived from independent sets of data. One was based on the $\chi_{O_3}^{\phi\lambda\tau}$ profile, which has a stratospheric maximum and a downward flux, and another was based on the $\overline{\chi}_{N_2O}^{\phi\lambda\tau}$ profile, which has a tropospheric maximum and an upward flux. The third was derived from the numerical tracer experiment data (Mahlman, 1972) of a general circulation model. For details of the derivations the reader is referred to the National Academy report (1979), *Stratospheric Ozone Depletion by Halocarbons: Chemistry and Transport.*

As shown in Figure 3-5 (A), each profile is different, but the profiles also have much in common. From a maximum at the surface they decrease monotonically to a minimum near 20 km and then increase in the upper stratosphere. There is no evidence of a discontinuity in $\partial K_z/\partial z$ corresponding to the tropopause, a feature introduced by many photochemical modelers, and the minimum is well above the height range of the tropopause, which varies from ~8 km at high latitudes to ~16 km near the equator.

The mean profile, denoted by a heavy continuous line in Figure 3-5 (A), ±33% includes all three. When multiplied by $\overline{\rho}$, the minimum shifts upward to ~24 km. For a constant vertical flux, this minimum represents a maximum resistance to vertical transport; consequently, $\partial \overline{\chi}_i^{\phi\lambda\tau}/\partial z$ achieves its maximum at about this level. Can we explain this profile from our current understanding of vertical transport processes? An interpretation proposed by this author identifies the minimum as the transition from a wave-dominated intrahemispheric regime, in which tropospheric generated waves damp with height, to an interhemispheric regime, in which the very long waves amplify with height. In the former, the maximum heating is near the equator with cooling near both poles. Heat is transported poleward in both hemispheres along slightly inclined paths. In the latter, the maximum heating is at the summer pole with the corresponding cooling at the winter pole. This differential heating drives a mean circulation in the upper stratosphere with rising motion over the heated pole, then a cross equatorial transport towards the winter pole where the air descends again. These two regimes are quite distinct, and the transition, when viewed from synoptic charts, varies from 20-23 km, in keeping with the $\overline{\rho}K_z$ minimum.

Gradients of diabatic heating and cooling, which provide the forcing function for both regimes, differ not just with latitude but also with

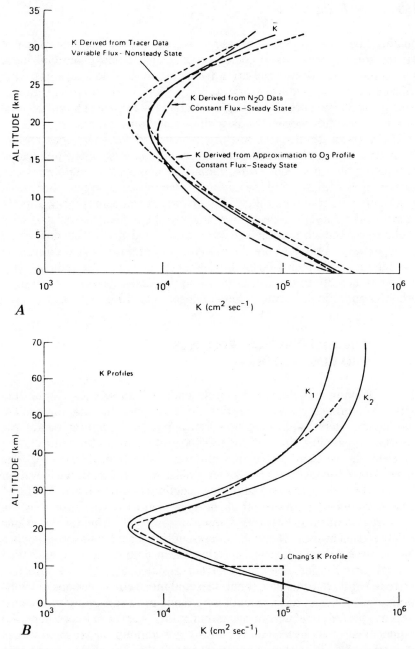

Figure 3-5. *(A):* Three derived K_z profiles (dashed lines) and mean profile (continuous line). *(B):* Extrapolated K_z profiles expressing lower and upper limits of uncertainty. *(From Danielsen, E. F., 1979, Derivation of K(z) profiles, in Stratospheric Ozone Depletion by Halocarbons: Chemistry and Transport, National Academy of Sciences, Washington, D.C., pp. 210, 213.)*

height. The lower region is heated at the surface by the absorption of visible solar radiation and cooled in the upper troposphere and lower stratosphere by infrared radiation from carbon dioxide and water. This differential heating reduces static stability, which leads to vertical mixing. The upward transport also is augmented by the release of latent heat as water vapor condenses in the ascending moist air currents.

The vertical gradient of heating is reversed for the upper region. Radiative cooling changes to heating, mainly because of the ultraviolet absorption by ozone. Since this gradient increases static stability, vertical mixing is inhibited and the mixing is predominantly quasi-horizontal, as a result of inertial instability. The net vertical transport is then directly linked to the diabatic processes: a downward transport associated with cooling over the winter pole and vice versa. Evidence of this downward transport was obtained from the movement of radioactivity produced by two high-altitude rocket-borne nuclear tests in 1958. The radioactivity moved seasonally to the winter pole and descended before spreading to other latitudes in the stratosphere (Telegadas and List 1964).

ONE-DIMENSIONAL PROFILES AND CORRELATIONS

Having established that independent data sets are compatible with the diffusion approximation to transport in one dimension, and having offered a physical interpretation of the K_z profile, we are in a position to examine the vertical transport of ozone and the correlation between the vertical profile of ozone and of potential vorticity. The method used to construct the ozone profile deserves special consideration because data are sparse and sampling is definitely nonuniform. The most extensive network of in situ profile measurements was established by the Air Force Cambridge Research Laboratory in 1963. It included stations widely spaced over the North American continent that made ozonesondes once a week or less frequently. Using mean data from 1963 and 1964 (Hering and Borden, 1965) the author grouped the stations into five latitude bands to construct zonal-seasonal mean cross-sections of ozone mixing ratio. Slight adjustments were made in the grouping to account for the expected zonal bias from sampling only over the one continent. An annual mean cross section was obtained by summing the seasonal values at a grid of points; then the approximation to the global mean was computed by integrating over all latitudes, with cos ϕ weighting to account for the decreasing area as ϕ increases from equator to pole.

The resulting profile of $\overline{\chi}_{O_3}^{\phi \lambda \tau}$ was quite smooth, but it seems likely that with extensive amounts of data it would be extremely smooth. Thus the final profile, shown in Figure 3-6, is the result of fitting analytic transition functions to the data. The hyperbolic function $T_K = e^x/(1+e^x)$, where $x = m_K (z - z_K)$, is a transition function centered at z_K. Its derivative $(1/m_K)(\partial T_K/\partial z)$ is a symmetric function also centered at z_K. The ozone profile was closely fitted by two transition functions. The nitrous oxide required a transition plus a symmetric function.

Since $\overline{\chi}_{O_3}^{\phi\lambda\tau}$ has a well-defined maximum at ~32 km, the diffusion approximation implies an upward flux above the maximum and downward flux below it. To test the sensitivity of the flux to the K_z profile two limiting profiles were constructed to bracket the range of uncertainty in K_z. The lower limit is expressed by K_1, the upper by K_2, and both were extended to 70 km by extrapolating the mean trend and assuming a maximum value at 70 km (see Fig. 3-5 (B). The resulting fluxes are shown in Figure 3-7 with the implied distributions of sources and sinks depicted in Figure 3-8. In both figures the dashed line represents the

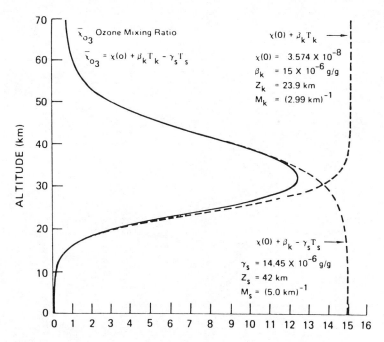

Figure 3-6. Ozone mixing ratio profile representative of global mean (continuous line) and analytic functions that describe it (dashed line). *(From Danielsen, E. F., 1979, Derivations of K(z) profiles, in Stratospheric Ozone Depletion by Halocarbons: Chemistry and Transport, National Academy of Sciences, Washington, D.C., p. 207.)*

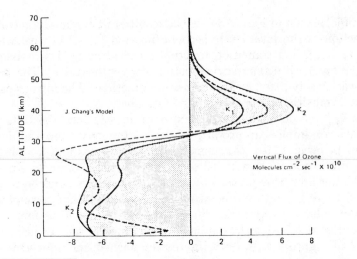

Figure 3-7. Vertical flux of ozone corresponding to K_1 (lower limit) and K_2 (upper limit). The dashed line is from J. Chang's numerical photochemical model. *(From Danielsen, E. F., 1979, Derivations of K(z) profiles, in Stratospheric Ozone Depletion by Halocarbons: Chemistry and Transport, National Academy of Sciences, Washington, D.C., p. 5.)*

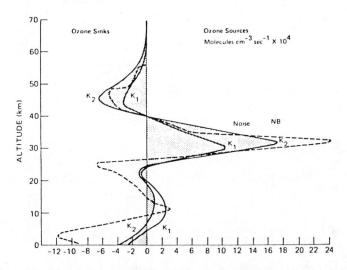

Figure 3-8. Ozone source–sink profiles derived from Figure 3-7. *(From Danielsen, E. F., 1979, Derivations of K(z) profiles in Stratospheric Ozone Depletion by Halocarbons: Chemistry and Transport, National Academy of Sciences, Washington, D.C., p. 5.)*

corresponding values from J. Chang's (private communication) chemical-diffusion model, which includes explicit ozone chemistry.

The results are qualitatively quite similar, differing most significantly in the magnitude of the tropospheric sources and sinks. From the major source at 32 km the upward flux reaches a maximum at 40 km; the downward flux increases in magnitude to ~25 km and then oscillates from 5 to 9×10^{10} molecules cm^{-2} s^{-1}. Another test was conducted with the NASA Ames Research Center's 1-D, stratospheric model (Turco and Whitten, (personal communication), which uses a constant flux at 10 km as a lower boundary condition. The average difference in $\overline{\chi}_{O_3}$ was 2% for the two K_z profiles, with the maximum at any level of 8%. Above 50 km the difference reduced to 0.3%.

On the basis of these results it would appear that the net vertical transport by vertical velocity deviations is statistically well simulated by the product of a diffusion coefficient and the vertical gradient of the mean mixing ratios. The analogy to molecular diffusion is apparently applicable because of the extremely large number of elemental volumes or masses in the atmosphere. Most satisfying is the conclusion that a single K_z profile can apply to an upward as well as a downward moving tracer.

If ozone has its major source in the middle stratosphere, we would expect transport to dominate in the lower stratosphere and upper troposphere. Under these circumstances there should be a positive correlation between $\overline{\chi}_{O_3}^{\phi\lambda\tau}$ and $\overline{S}^{\phi\lambda\tau}$. The profile of $\overline{S}^{\phi\lambda\tau}$ shown in Figure 3-9 is

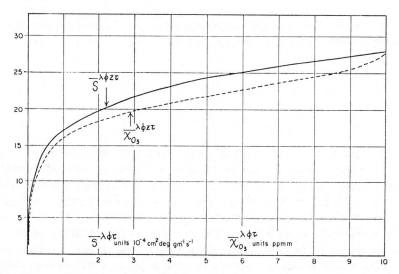

Figure 3-9. Northern hemispheric mean profile of potential vorticity (solid line) and corresponding ozone mixing ratio profile (dashed line). Arrows denote respective vertical means from surface to 28 km.

remarkably similar to that of $\overline{\chi}_{O_3}^{\phi\lambda\tau}$, and the correlation is $+\,0.974$ from the surface to 28 km. This large value definitely supports the assumption that transport dominates photochemical processes as ozone moves downward towards the surface.

The $\overline{S}^{\phi\lambda\tau}$ profile was computed as was $\overline{\chi}^{\phi\lambda\tau}$ by integrating $\overline{S}^{\lambda\tau}$ cos ϕ with respect to latitude. However, $\overline{S}^{\lambda\tau}$ is derived directly from northern hemispheric analyses of $\overline{z}^\tau$ and $\overline{T}^\tau$ at the mandatory pressure surfaces. Here the $(\overline{\quad})^\tau$ denotes a monthly mean. All phases of the analysis are objectively determined, including initial error detection at each radiosonde station, computation of the local monthly mean values, grid point analysis, filtering of the analyses by double Fourier decomposition, zonal averaging of $\overline{z}^\tau$ and $\overline{T}^\tau$, computation of $\overline{U}^{\lambda\tau}$ and $\overline{\theta}^{\lambda\tau}$, and finally from these variables the computation of $\overline{S}^{\lambda\tau}$. To assure representative values, $\overline{S}^{\lambda\tau}$ was computed for the central month of each season and for two pairs of successive years, 1973-1974 and 1979-1980. The latter pairing eliminates the differences in $\overline{U}^{\lambda\tau}$ and $\partial\overline{\theta}^{\lambda\tau}/\partial z$ associated with the biannual oscillations. Thus, $\overline{S}^{\lambda\tau}$ and $\overline{S}^{\phi\lambda\tau}$ are highly representative of the earth's atmospheres for the northern hemisphere, averaged, of course, over all seasons of the year. We believe they are representative also of 1963-1964 when the ozone measurements were made.

TWO-DIMENSIONAL TRANSPORT, MERIDIONAL AND VERTICAL FLUXES

With elimination of latitudinal averaging, profiles become distributions and transport is augmented by advection by mean motions. Also, there are two fluxes,

$$\overline{F}_y^{\lambda\tau} = \overline{\rho v \chi}_{O_3}^{\lambda\tau} = \overline{\rho}\ \overline{v}^{\lambda\tau}\overline{\chi}_{O_3}^{\lambda\tau} + \overline{\rho}\ \overline{v'\chi'_{O_3}}^{\lambda\tau}, \tag{3-10}$$

$$\overline{F}_z^{\lambda\tau} = \overline{\rho w \chi}_{O_3}^{\lambda\tau} = \overline{\rho}\ \overline{w}^{\lambda\tau}\overline{\chi}_{O_3}^{\lambda\tau} + \overline{\rho}\ \overline{w'\chi'_{O_3}}^{\lambda\tau}. \tag{3-11}$$

The expansions on the right side are traditionally made because $\overline{v}^{\lambda\tau}$ can be determined from the observations and $\overline{w}^{\lambda\tau}$ can be derived from the mass continuity equation. Presumably this leaves only the deviatory fluxes to be parameterized in terms of mean mixing ratio gradients. Unfortunately, the expansion terms are not independent when the atmosphere contains waves whose v and w velocities are circularly or elliptically polarized. Consider a large scale wave whose air parcel trajectories oscillate north-south and vertically with a phase shift between them, such that they produce a closed elliptical circulation when projected on a meridional-vertical plane. After a complete wave period an air parcel could travel several thousand kilometers eastward but its projection on the plane would return to the same initial point. With zonal-temporal averaging over one or more complete periods the net displacements of all parcels are zero as is the net transport.

Despite this obvious physical constraint, these waves will contribute to $\bar{v}^{\lambda\tau}$ and $\bar{w}^{\lambda\tau}$ and hence to mean advective transports whenever their amplitudes vary with ϕ and z. Since these waves do not produce a net transport, their advective transport must be cancelled by an equal and opposite deviatory transport. The effect was demonstrated by Mahlman (1972) from the results of a tracer experiment performed with a numerical general circulation model. He showed that the net flux was often a small residual of a large mean flux and an opposing deviatory flux.

The above arguments are valid if and only if the wave-induced motions are isentropic, that is, adiabatic and reversible. We used the adiabatic assumption to complete trajectories in Figure 3-1 (*C*). The projections of these trajectories on a meridional-vertical plane would trace out families of ellipses with their major axis sloping upward to the north. The circulations would be thermally indirect as in the Ferrel cell, a common feature of Eulerian averaging in extratropical latitudes. However, we deduced from these adiabatic trajectories that the flow would not be reversible due to a combination of wave amplification, mixing, and diabatic heating. In fact, these amplifying, elliptially polarized waves are responsible for the major transport of lower stratospheric air, including ozone, into the troposphere.

The example illustrates the complexity of the problem. Zonally averaged mean motions are combinations of material flows and streamlines which in a Fourier decomposition represent the transport by longitudinal wave number zero. In a complex atmosphere, such as the earth's, they have no simple physical significance until they are coupled with the deviatory motions. The latter include the effects of all nonzero wave numbers whose perturbation velocities can be orders of magnitude larger than the small residual mean velocities.

During the last decade considerable effort has been directed towards determining a zonal average not at a constant latitude but along an undulating material flow line or tube to eliminate or reduce the deviatory motions, in particular the isentropic elliptical motions that lead to opposing fluxes between the Eulerian mean and deviatory terms. Unfortunately, this concept, so appealing when applied to a single small-amplitude quasi-steady wave, becomes unworkable when applied to the earth's atmosphere with its broad spectrum of finite-amplitude waves and intermittent instabilities. Perhaps the most disturbing consequence of this effort has been general acceptance of the concept that transport could be reduced to only advection by properly corrected mean motions.

Here we must recall that advection is a nondispersive transport and stress that we have considerable evidence that atmospheric transport is highly dispersive. The best examples are the dispersion of radioactive isotopes from nuclear bomb tests in the atmosphere and the dispersion of volcanic eruption plumes. In these examples the sources are isolated and their initial concentrations and masses can be reasonably estimated. As the concentrations or mixing ratios decrease with time the mass of air into which they mix must simultaneously increase. In a recent study by the author (unpublished) of the rate of decrease of radioactive tungsten from the 1958 Hardtack series, based on aircraft sampling, the mass of air increased by five orders of magnitude in a period of only three months. During this time it spread from its equatorial stratospheric injection source into the stratosphere of both hemispheres. The characteristic mixing times increased from about 10^5s to 10^6s.

This result supports the numerical modeling assumptions made by Davidson, Friend, and Seitz (1966) who neglected the mean advection terms in (3-10) and (3-11) and parameterized the deviatory fluxes by a symmetric transport tensor to simulate the transport and fallout of radioactivity. It also supports Reed and German's (1965) use of a purely diffusive analytical model to simulate the major feature of the tungsten transport. In both models the parameterizations were an extension of the assumptions used in 1-D models. Thus,

$$\overline{F}_y^{\lambda\tau} = -\overline{\rho} \left[K_{yy} \frac{\partial \overline{X}^{\lambda\tau}}{\partial y} + K_{yz} \frac{\partial \overline{X}^{\lambda\tau}}{\partial z} \right], \tag{3-12}$$

$$\overline{F}_z^{\lambda\tau} = -\overline{\rho} \left[K_{zy} \frac{\partial \overline{X}^{\lambda\tau}}{\partial y} + K_{zz} \frac{\partial \overline{X}^{\lambda\tau}}{\partial z} \right]. \tag{3-13}$$

Since advection by mean motions was neglected, an additional constraint was required, that $K_{zy} = K_{yz}$. The reason for this constraint will be discussed after we consider the implications of their model simulations for ozone transport, specifically for zonal-temporal mean transport in the lower stratosphere.

Several important conclusions can be drawn from these studies.

1. Transport in the lower stratosphere is inclined, not horizontal.
2. Meridional flux is one thousand times the vertical flux consistent with the ratio of horizontal to vertical wind speeds.
3. $K_{yy} \simeq 10^3 \mid K_{yz} \mid$ and $\mid K_{yz} \mid \simeq 10^3 K_{zz}$. These ratios imply $K_{yy} \simeq \overline{vv}\tau$, $K_{yz} \simeq \overline{vw}\tau$, and $K_{zz} \simeq \overline{ww}\tau$, where τ is a representative mixing time.
4. Transport by deviatory motions dominates transport by zonal-temporal mean motions.
5. Dispersion by deviatory motions dominates advection by deviatory motions during the first two months after injection.

Reed and German (1965) characterized their purely dispersive model, with constant K_{yy}, K_{yz}, and K_{zz}, as a crude simulation to illustrate the usefulness of the concepts. Indeed, a comparison of their computed and observed distributions, their Figure 3, shows significant deviations in tropical latitudes, indicating that advective transport cannot be entirely neglected. A more realistic distribution was achieved by Louis (1974), who first derived representative Eulerian mean velocities and then applied reasonable constraints to evaluate the K's. In the lower stratosphere he assumed that the ozone mixing ratios were conserved and in steady state. He then derived K's that would cancel the local changes due to advection by mean motions. The resulting distributions of the K's included spatial gradients that produce advections by the deviatory motions. It should be noted that in the predictive ozone continuity equation it is the divergence of the fluxes F_y and F_z that determine the local change. Spatial gradients of the K's times the ozone mixing ratio gradients then simulate advective transports while the products of the K's and the second derivatives of the mixing ratio simulate dispersion.

Louis's model was used to simulate the transport of radioactive debris from seveal Chinese nuclear bomb tests between 1968 and 1972 as well as the tungsten-185 from the U.S. Hardtack tests, 1958. It reproduced quite accurately the decrease in stratospheric burden, the distributions in the stratosphere, and the latitudinal variation in fallout. It also reproduced the observed distribution of tungsten at tropical latitudes, including the upward and equatorward movement of the center of maximum mixing ratio. The latter requires an advective transport. In these independent tests based on aircraft observations below 21 km the model performed well, justifying the assumption that ozone is transported as if it were a passive tracer in the lower stratosphere. However, when the transport was combined with photochemistry (Crutzen, private communication) the poleward transport of ozone in the middle stratosphere was underestimated.

At altitudes above 21 km we lack in situ measurements of passive tracers to determine the relative importance of the mean and deviatory transports. It is quite probable that the diabatically driven interhemispheric mean circulations play an important role in the upper stratosphere, but in the middle stratosphere, where K_z is a minimum, advection and dispersion by deviatory motions are probably dominant. This hypothesis is supported by the large meridional displacements and mixing determined by trajectories computed at these altitudes. The trajectories frequently include large anticyclonic loops indicative of inertial instability (Danielsen, 1981a).

In the future it should be possible to use potential vorticity as a passive tracer by combining in situ radiosonde observations with remote satellite measurements. If representative objective analyses of u, v and $\overline{S}$ can be achieved, w can be computed from them and the method proposed by

Danielsen (1981*b*) can be used to derive $\overline{v}^{\lambda\tau}$, $\overline{w}^{\lambda\tau}$, K_{yy}, K_{yz}, K_{zy}, and K_{zz}. Accepting Matsuno's (1980) concept of a mixing time, the method reduces the transport problem to two equations in two unknowns, the weighting functions for the components of an antisymmetric and symmetric transport tensor. When expressed in these terms, (3-10) and (3-11) can be written as a vector equation,

$$\mathbf{F}^{\lambda\tau}(O_3) = \overline{\rho}\overline{\mathbf{V}}^{\lambda\tau}\,\overline{X}^{\lambda\tau}_{O_3} - \overline{\rho}\mathfrak{K}_A \cdot \nabla \overline{X}^{\lambda\tau}_{O_3} - \overline{\rho}\mathfrak{K}_S \cdot \nabla \overline{X}^{\lambda\tau}_{O_3}, \tag{3-14}$$

where the antisymmetric tensor expressed in matrix form is

$$[\mathfrak{K}_A] = \begin{bmatrix} 0 & K_A \\ -K_A & 0 \end{bmatrix} \tag{3-15}$$

and the symmetric tensor is

$$[\mathfrak{K}_S] = \begin{bmatrix} K_{yy} & K_S \\ K_S & K_{zz} \end{bmatrix}. \tag{3-16}$$

In (3-15) and (3-16)

$$K_S = \frac{K_{yz} + K_{zy}}{2} \tag{3-17}$$

and

$$K_A = \frac{K_{yz} - K_{zy}}{2}. \tag{3-18}$$

The second term on the right of (3-14), called the *Stoke's correction*, cancels the nonphysical component of the Eulerian mean motions due to isentropic, elliptically polarized deviatory motions. Thus the sum of the first and second terms is the physically meaningful mean flux. Because

$$\nabla \cdot \overline{\rho}\,\overline{\mathbf{V}}^{\lambda\tau}\,\overline{X}^{\lambda\tau}_{O_3} = \overline{\rho}\,\overline{\mathbf{V}}^{\lambda\tau} \cdot \nabla \overline{X}^{\,\lambda\tau}_{O_3} + \overline{X}^{\lambda\tau}_{O_3}\overline{\nabla \cdot \overline{\rho}\,\overline{\mathbf{V}}^{\lambda\tau}}^{\,0}$$

the vanishing mass divergence indicates that the vector $\overline{\rho}\,\overline{\mathbf{V}}^{\lambda\tau}$ can be expressed by the gradient of a mass stream function. Therefore, the first and second terms in (3-14) can be written as antisymmetric tensors and combined as one. Then (3-14) reduces to

$$\mathbf{F}^{\lambda\tau}(O_3) = \mathfrak{K}^*_A \cdot \nabla \overline{X}^{\lambda\tau}_{O_3} - \overline{\rho}\mathfrak{K}_s \cdot \nabla \overline{X}^{\lambda\tau}_{O_3}, \tag{3-19}$$

where K_A is the stream function for the corrected mean mass flux. Here we see why neglecting all mean motions requires the symmetric constraint $K_{zy} = K_{yz}$. Physically, it implies that all deviatory wave-induced motions are linearly polarized, that is, that v and w osillate in phase so that the trajectories lie in inclined planes.

DESCRIPTION OF TWO-DIMENSIONAL TRANSPORT

When the antisymmetric and symmetric tensors are included in the generalized transport tensor as in (3-19), the projected motions describe inclined ellipses. In the lower stratosphere they slope downward toward the pole, while in the extratropical troposphere they slope upward toward the pole. Thus, in the mean, ozone is transported from its predominately tropical stratospheric source region into the lower polar stratosphere. From there it is intermittently transferred into the troposphere along inclined paths toward lower latitudes. Because the transport processes in the lower stratosphere and troposphere are directly linked to the extratropical cyclones, the amount of ozone transferred varies seasonally as do the preferred latitudes and longitudes of transfer.

The cyclonic storms are generally weaker and farther north in the summer, stronger and farther south in the winter. They intensify most rapidly where the static stability is small, as a result of surface heating. In the winter, cold air is destabilized most rapidly when it moves over warmer oceans. In the spring, the cyclogenesis shifts to the heated arid continents. These longitudinal shifts vanish, of course, with zonal-temporal averaging, but the seasonal dependence on latitude remains if the temporal average is limited to times shorter than a season. The seasonal variation in amount of ozone transferred remains also. It reaches a maximum in late spring when larger mixing ratios have reached the lower stratosphere and when the mass of air in the lower stratosphere itself is most rapidly reduced. Physically, the volume of the stratosphere contracts as the storms retreat northward with the approach of summer.

It expands again in the fall when more air is transported into the stratosphere than is removed from it. The influx from troposphere to stratosphere remains to be quantified and the processes specifically identified. Probably, the major influx occurs at low latitudes as a consequence of the large convective clouds that form in the convectively unstable lower tropical troposphere.

TWO-DIMENSIONAL DISTRIBUTIONS AND CORRELATIONS

The distribution of $\overline{\chi}_{O_3}^{\lambda\tau}$ versus latitude and height (Fig. 3-10) is contoured in parts per million by mass. To emphasize the similarity in

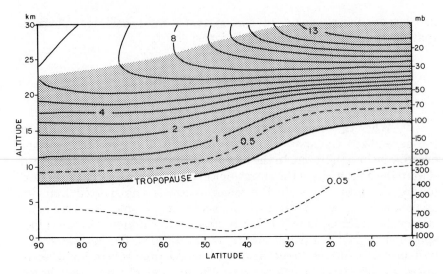

Figure 3-10. Distribution of zonal–annual mean of ozone mixing ratio in ppmm. Toned area extends from mean tropopause to axis of relative maximum.

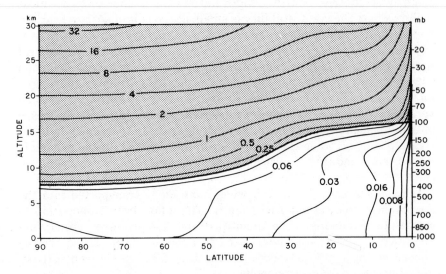

Figure 3-11. Distribution of zonal-annual mean of potential vorticity contoured in units of 10^{-4} cm^2 deg gm^{-1} s^{-1}. Toned area denotes mean stratosphere.

slope of the isopleths in the lower stratosphere to the mean tropopause, a grey toned transition layer ~15 km deep has been included. Its lower boundary identifies the mean tropopause; its upper boundary coincides with the axis of relative maxima that extends from the absolute tropical maximum. For direct comparison with $\overline{\chi}_{O_3}^{\lambda\tau}$, the distribution of $\overline{S}^{\lambda\tau}$ is presented in Figure 3-11. In this figure, adjacent contours differ by a factor of two to resolve the much larger range of values.

The two distributions are quite similar in the lower stratosphere, but they are definitely dissimilar above the toned transition layer in Figure 3-10. To quantify these conditions, we have computed the area-weighted correlation between χ'_{O_3} and S'. The deviations, defined by

$$(\quad)' = (\overline{\qquad})^{\lambda\tau} - (\overline{\qquad})^{\phi\lambda\tau} \qquad (3\text{-}20)$$

were multiplied, and their product, weighted by cos ϕ, was integrated from 5N to the pole. The results (Fig. 3-12) are extremely interesting. From a negative correlation between 28 and 26 km there is a rapid change to a positive correlation. The positive correlation is maintained close to 1 from 22 km to 8 km; then it decreases in magnitude but remains positive in the lower troposphere.

Since the diabatic heating rates and their vertical gradients are large in

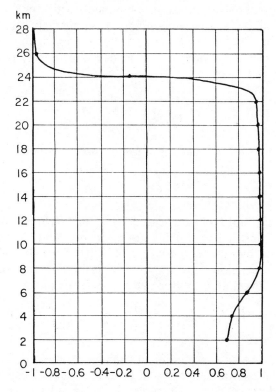

Figure 3-12. Area-weighted correlation between ozone mixing ratio and potential vorticity deviations. See text for definition of deviations. Negative values imply photochemical dominance, positive imply transport dominance.

the upper stratosphere but small in the middle stratosphere, potential vorticity is quasi-conserved within the height range of Figure 3-12. The negative correlation between 28 km and 26 km reflects the higher altitude, higher latitude source for potential vorticity and the low latitude source for ozone. Mixing is the only reasonable explanation for the change from negative to positive correlations. After $\overline{S}^{\lambda\tau}$ and $\overline{\chi}_{O_3}^{\lambda\tau}$ are highly correlated, mixing will not destroy the correlation; thus both quantities decrease in magnitude with decreasing height.

We can conclude from the correlation profile that if significant amounts of ozone are produced in the troposphere they are also destroyed there or at the earth's surface. There certainly is no evidence of a decrease in the positive correlation between 16 and 8 km, the height range of the mean tropopause from tropics to pole. We know that ozone is produced during the sunlit hours in polluted surface boundary layers. This ozone will be negatively correlated with potential vorticity. If it were transported upward it would reduce the correlations in the 8–16 km range, where on the basis of our present data the positive correlations are a maximum. Perhaps more extensive tropical measurements will alter this conclusion.

Until they do, we can take advantage of the unusually high correlations to estimate the effects of transport on the meridional variations of total ozone. If we use the symbol $T(O_3)$ to denote the total mass of ozone overhead per unit area, then

$$T(O_3) = \frac{M_{O_3}}{A} = \frac{\iiint \rho_{O_3}\, dV}{\iint dA} = \int_0^\infty \chi_{O_3}\rho\, dz = \frac{\overline{\chi}_{O_3} M}{A}, \qquad (3\text{-}21)$$

where M is the mass of air above the same area. Similarly, we can define

$$T(S) = \int_0^\infty S\rho\, dz = \frac{\overline{S}M}{A}. \qquad (3\text{-}22)$$

Therefore, if χ_{O_3} and S are correlated, $T(O_3)$ and $T(S)$ will be correlated, and in particular, if $\overline{\chi}_{O_3}^{\lambda\tau}$ and $\overline{S}^{\lambda\tau}$ are correlated and $\overline{\chi}_{O_3}^{\phi\lambda\tau}$ and $\overline{S}^{\phi\lambda\tau}$ are correlated, the meridional variation of $\overline{T}^{\lambda\tau}(O_3)$ will be related to the meridional variation in $\overline{T}^{\lambda\tau}(S)$. By substituting (3-6) into (3-22), we can simplify it to reveal an important atmospheric subdivision:

$$\overline{T}^{\lambda\tau}(S) = \int_0^\infty \overline{S}^{\lambda\tau}\rho\, dz = \int_0^\infty \overline{\alpha\frac{\partial\overline{\theta}^{\lambda\tau}}{\partial z}}\,\overline{\eta}_\theta^{\lambda\tau}\rho\, dz = \int_{\theta 0}^{\theta\infty} \overline{\eta}_\theta^{\lambda\tau}\, d\overline{\theta}^{\lambda\tau}. \qquad (3\text{-}23)$$

From (3-23) we can conclude that when $\overline{\chi}_{O_3}^{\lambda\tau}$ and $\overline{S}^{\lambda\tau}$ are positively correlated, the meridional variation in total ozone will follow the variation in absolute vorticity. To relate the latter to the atmosphere we turn to Figure 3-13. It depicts the meridional vertical distribution of

$\overline{\theta}^{\lambda\tau}$ and $\overline{U}^{\lambda\tau}$, the mean zonal wind speed. Recalling that $\overline{\eta}_{\theta}^{\lambda\tau} = f - \partial\overline{U}^{\lambda\tau}/\partial y$ $= f - (1/R)\,\partial\overline{U}^{\lambda\tau}/\partial\phi$, we see immediately that $\overline{\eta}_{\theta}^{\lambda\tau}$ is large at high latitudes because f is large and $-\partial\overline{U}^{\lambda\tau}/\partial y > 0$. Conversely, at low latitudes f is small and $-\partial\overline{U}^{\lambda\tau}/\partial y < 0$. In Figure 3-13, the mean shear vorticity changes sign where $\overline{U}^{\lambda\tau}$ is a maximum, about 20°N of the $\overline{U}^{\lambda\tau}$ $= 0$ isoline. Thus the maximum positive $\overline{U}^{\lambda\tau}$ at each constant θ surface, the relative jet at constant θ, meridionally separates large from small inertial stability just as the conventional tropopause vertically separates large from small static stability. If we make both separations, the atmosphere is subdivided into quadrants as shown schematically in Figure 3-14.

Because $\overline{a}$, $\partial\overline{\theta}^{\lambda\tau}/\partial z$, and $\overline{\eta}_{\theta}^{\lambda\tau}$ are each larger in the cyclonic stratosphere than in the anticyclonic troposphere, their product $\overline{S}^{\lambda\tau}$ is much larger in the former than in the latter. The ozone distribution involves one additional complexity: $\partial\overline{\chi}_{O_3}^{\lambda\tau}/\partial y$ changes sign near 23 km where the $\overline{\chi}_{O_3}^{\lambda\tau}$ and $\overline{S}^{\lambda\tau}$ correlation also changes sign. To include this sign reversal we have introduced another subdivision into the stratosphere. Thus Figure 3-15 depicts the major subdivisions and their contributions to total ozone and to the total density weighted potential vorticity. The diagram extends to 29 km or to $\sim$15 mb. The amount of ozone between p = 0 and p = 15 mb, that which does not appear in Figure 3-15, probably is $<0.8 \times 10^{-4}$ gm cm^{-2}.

From Figure 3-15 we note that the major contribution to the meridio-

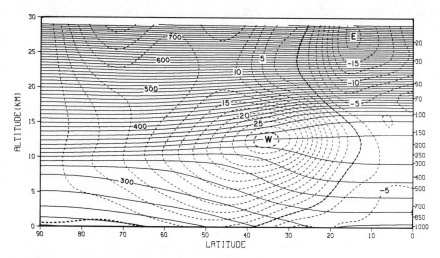

Figure 3-13. Zonal–biannual mean of potential temperature $\overline{\theta}^{\lambda\tau}$ and zonal wind velocity $\overline{U}^{\lambda\tau}$ (dashed lines every 2.5 m s^{-1}). **W** denotes maximum winds from the west, **E** from the east.

nal variation in total ozone is from the cyclonic and anticyclonic stratosphere below 23 km. The cyclonic stratosphere contributes >4 times the anticyclonic stratosphere, with about a factor of two due to the difference in mass of air in the two regions and another factor of two due to the larger ozone mixing ratios in the cyclonic stratosphere.

This meridional ratio of 4/1 is reduced to <2/1 by the compensating gradients in the upper stratosphere. The tropospheric gradient is extremely small, but adding the same amount to the central layer changes the ratio to 3/1. Below 23 km the ratios of relative contributions to total ozone are almost identical to the ratios of the mass weighted potential vorticity. These results imply that the total ozone is much larger poleward than equatorward of the jet and that transport dominates photochemistry in the mean meridional gradient of total ozone.

DIFFERENTIAL TRANSPORT AND SYNOPTIC DISTRIBUTIONS

Synoptic distributions of ozone mixing ratio differ from mean distributions by being folded, that is, they have reversals of gradients with relative maxima and minima. With few exceptions, space-time continuity of these relative maxima or minima cannot be determined from the widely spaced ozone measurements. Instead we must rely on correlations with other trace measurements or more frequently on correla-

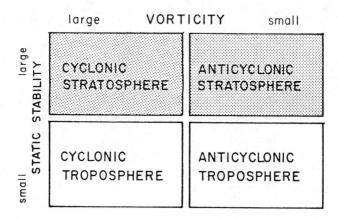

Figure 3-14. Schematic subdivisions of atmosphere according to static and inertial stability (vorticity).

tions with simultaneous meteorological measurements to resolve the ozone distribution.

Having conducted several field experiments with multiple aircraft equipped to measure meteorological variables, trace gases, and aerosols, this author has found that most variations are produced by differential transport and hence can be identified by cross correlations with other variables. On the large scale, differential transport tends to fold horizontal gradients to produce relative maxima and minima in the vertical. For

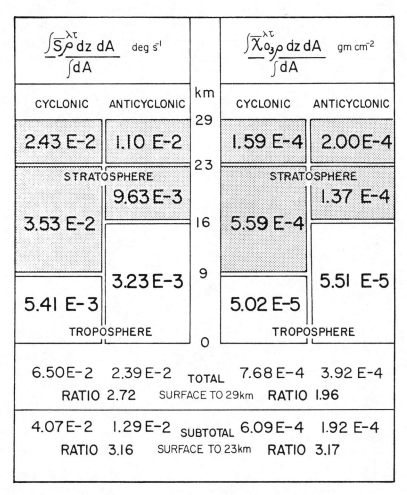

Figure 3-15. Contributions of dynamic subdivisions to total ozone and total potential vorticity.

example, ozone-rich air from the lower cyclonic stratosphere can be advected southward and downward into the troposphere generating a relative maximum of ozone in the troposphere. This maximum has spatial continuity, as shown schematically in Figure 3-4, and in an actual potential vorticity distribution, Figure 3-16 (*B*).

The vertical cross section of $\overline{\theta}$ and $\overline{U}$, Figure 3-16 (*A*), shows a stable-layer large $\partial\theta/\partial z$ extending from the lower stratosphere over Winnemucca, Nevada to the unstable dust-laden boundary layer over Arizona and New Mexico. That the layer is extruded from the stratosphere can be seen from the potential vorticity and the radioactivity. Notice that the layer is extruded beneath the core of the polar jet and that a smaller layer extends beneath the core of the subtropical jet near Winslow, Arizona. Note also that the potential vorticity and radioactivity decrease

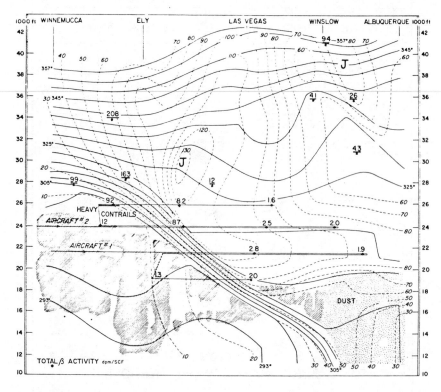

Figure 3-16 *(A).* Vertical cross-section from Project Springfield of potential temperature $\overline{\theta}$ and wind speed $\overline{U}$. Total β activities were measured from filters collecting radioactive aerosols on upwind–downwind flights. *(From Danielsen, E. F., 1964, Report on Project Springfield, DASA 1517, Defense Atomic Support Agency, Washington, D.C.)*

down the extruded layer due to mixing with the tropospheric air. However, the large scale motions are frontogenetical: convergence counteracts the dispersion normal to the layer while divergence in the plane of the layer augments it, spreading the decreasing concentrations to a larger and larger area.

Correlations between ozone mixing ratio, radioactivity, and potential vorticity from another aircraft experiment are shown in Figure 3-17. In this case three RB-57 aircraft equipped to take bulk filter samples for radiochemical analyses were flown at 29, 37, and 45 thousand feet, where they made upwind-downwind spot samples. At the same time a fourth aircraft equipped with a Komhyr type ozone analyzer was flown cross-wind at the same respective altitudes and at 24,000 feet. As is evident in

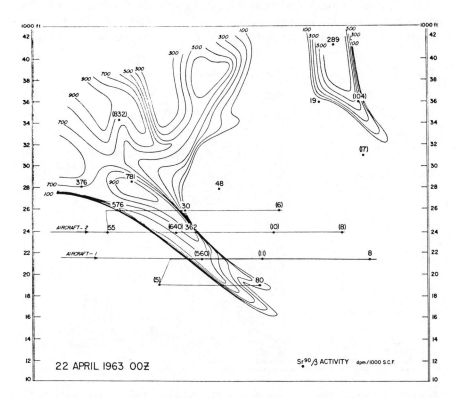

Figure 3-16 *(B)*. Vertical cross-section of potential vorticity $\overline{S}$ showing folded tropopause structure. $\overline{S}$ contoured in units of 10^{-7} cm^2 deg gm^{-1} s^{-1} to permit direct numerical comparison to ^{90}Sr activity. *(From Danielsen, E. F., 1964, Report on Project Springfield, DASA 1517, Defense Atomic Support Agency, Washington, D.C.)*

Figure 3-17, the ozone and radioactivity show a triple fold where the potential vorticity indicates a single fold.

Instantaneous measurements of radioactivity from the other flight, Figure 3-16, did not show this multiple structure except near the southern limit. So we conclude that the two folded structures of Figures 3-16 and 3-17 are different in that respect. However, due to the wide separation of radiosonde observations, we cannot resolve potential vorticity variations on the scale of the observed ozone variations. To correlate ozone and potential vorticity we would first filter the aircraft measurements of ozone with a low-pass filter to remove the short wavelength oscillations.

The large scale transport, as discussed earlier, is quasi-horizontal, but it leads to large vertical displacements because the vertical motions are sustained for periods of 1-3 days. There is, of course, a broad spectrum of much shorter-period waves that effectively spreads the ozone into the surrounding air. These wave modes can be isolated and identified by using band-pass filters on data obtained by a fast responding ozone

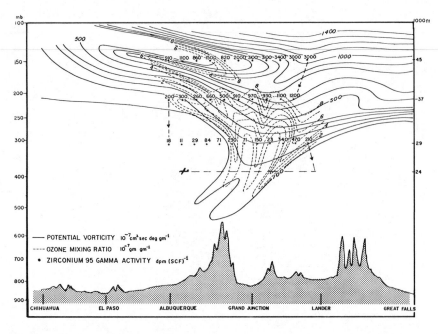

Figure 3-17. Vertical cross-section of potential vorticity (contoured here as $\bar{S} \triangleright^{-1}$), ozone mixing ratio, and gamma activity of zirconium-95. *(From Danielsen, E. F., R. Bleck, J. P. Shedlovsky, A. Wartburg, P. Haagenson, and W. Pollack, 1970, Observed distribution of radioactivity, ozone and potential vorticity associated with tropopause folding, J. Geophys. Res. **75;** copyright © 1970 by the American Geophysical Union.)*

sensor. In addition, having extracted the wave amplitudes and phase, one can compute the horizontal and vertical fluxes of ozone, potential temperature, horizontal momentum, and so on, as functions of time.

We have made calculations of this type from aircraft measurements made as the NCAR Electra traversed the folded tropopause described by Danielsen and Mohnen (1977). The Electra traversed the layer four times. The potential temperature and momentum fluxes were negative for the sum of the four traverses, but for three of the four the fluxes were positive. During part of the remaining traverse, large-amplitude waves with periods relative to the aircraft of ~ 1 min were encountered for approximately 10 min. The flux during this interval was so strongly negative that it dominated the average of the four traverses, which extended over 100 min.

We stress this point here because flux computations are now possible with the fast responding instruments, inertial navigation, and air motion sensors. They are becoming increasingly popular, also. We caution that the smaller-scale fluxes are intermittent in time and inhomogeneous in space, as are the large-scale fluxes and transports. Consequently, care must be taken to assure representative statistics by accumulating enough samples, before reasonable conclusions can be drawn.

SUMMARY

In the upper stratosphere, 30-50 km, ozone is in photochemical equilibrium; thus the importance of transport is relatively insignificant. For what it is worth, the transport is dominated by an interhemispheric circulation with flow from the southern to the northern hemisphere and descent at the north pole in December, January and February. Six months later the flow reverses and the air descends near the south pole.

Below 30 km the importance of transport increases as does the importance to transport of deviatory motions relative to mean motions. Direct evidence of both is the change from negative to positive correlation between ozone mixing ratio and potential vorticity, a change that can only be attributed to transport in general and to deviatory motions in particular. Between 25 km and the tropopause, ozone is transported poleward and downward, as is the potential vorticity, to the cyclonic side of the subtropical and polar jets. Then it intermittently leaves the lower cyclonic stratosphere, as the tropopause folds during large scale cyclogenesis, flowing equatorward and downward toward the surface boundary layer.

The large positive correlation, >0.95, between ozone mixing ratio and potential vorticity deviations in the lower stratosphere and upper troposphere assures us that transport alone will explain the seasonal variations of ozone with respect to latitude and longitude within the same height range, 22-8 km, and also the corresponding variations in total ozone. In addition, it assures us that the net transport is downward from the upper-middle stratosphere to the earth's surface. If a significant ozone source exists in the troposphere, this ozone must also be destroyed at the earth's surface, because there is no evidence of a negative correlation between ozone mixing ratio and potential vorticity deviations in the upper troposphere. A negative correlation would be a definite indicator of ozone production in the troposphere.

Finally, the predominance of adiabatic mixing, whi ⅃ is irreversible and dispersive, indicates that nondispersive advective processes, whether Eulerian or Lagrangian in their formulation, can not describe the ozone transport. This statement does not imply that the contributions to advection by diabatic processes are unimportant. They are essential, but they are not sufficient.

POSTSCRIPT

A multiaircraft experiment directed by the author to study transport of trace gases and aerosols during tropospheric folding events was completed recently. Preliminary results definitely confirm the bulk transport of ozone from the lower stratosphere through the troposphere to the surface boundary layer. They also support the positive correlation between potential vorticity and the negative correlation with carbon monoxide mixing fractions.

ACKNOWLEDGMENT

The author thanks R. Stephan Hipskind and Steven E. Gaines for programming the objective analysis methods and computer graphics used to produce the zonal-monthly and zonal-annual means of potential temperature, zonal wind velocity, and potential vorticity.

REFERENCES

Danielsen, E. F., 1961, Trajectories: Isobaric, isentropic and actual, *J. Meteorol.* **18**:479-486.

Danielsen, E. F., 1964, *Report on Project Springfield*, DASA 1517, Defense Atomic Support Agency, Washington, D.C.

Danielsen, E. F., 1968, Stratospheric-tropospheric exchange based on radio activity, ozone and potential vorticity, *J. Atmos. Sci.* **25**:502-518.

Danielsen, E. F., 1974, Review of trajectory methods, *Adv. Geophys.* **18**:73-94.

Danielsen, E. F., 1979, Derivations of K(z) profiles, *Stratospheric Ozone Depletion by Halocarbons: Chemistry and Transport*, National Academy of Sciences, Washington, D.C., 202-218.

Danielsen, E. F., 1981*a*, Trajectories of the Mount St. Helens eruption plume, *Science* 211:819-821.

Danielsen, E. F., 1981*b*, An objective method for determining the generalized transport tensor for two-dimensional Eulerian models, *J. Atmos. Sci.* 38:1319-1339.

Danielsen, E. F., and V. A. Mohnen, 1977, Project Dustorm Report: Ozone transport, in situ measurements and meteorological analyses of tropopause folding, *J. Geophys. Res.* 82:5867-5877.

Danielsen, E. F., R. Bleck, R. Adler, D. Deaven, C. Dey, V. Kousky, and F. Sechrist, 1966, *Research in Four-Dimensional Diagnosis of Cyclonic Storm Cloud Systems*, AFCRL-66-30, Air Force Cambridge Research Laboratory Rep. No. 2, Hanscom Field, Mass.

Danielsen, E. F., R. Bleck, J. P. Shedlovsky, A. Wartburg, P. Haagenson, and W. Pollack, 1970, Observed distribution of radioactivity, ozone and potential vorticity associated with tropopause folding, *J. Geophys. Res.* 75:2353-2361.

Davidson, B., J. P. Friend, and H. Seitz, 1966, Numerical models of diffusion and rainout of stratospheric radioactive materials, *Tellus* 18:301-315.

Ertel, H., 1942, Ein neuer hydrodynamischer Wirbelsatz, *Meteorl. Z.* 59:277-281.

Hering, W. S., and T. R. Borden, Jr., 1965, *Mean Distributions of Ozone Density over North America, 1963-1964*, AFCRL-65-913, Air Force Cambridge Research Laboratory, Hanscom Field, Mass.

Louis, J. F., 1974, A Two-Dimensional Transport Model of the Atmosphere, Ph.D. thesis, University of Colorado, Boulder.

Mahlman, J. D., 1972, Preliminary results from a three-dimensional, general-circulation/tracer model, in *Proceedings of the Second Conference on the Climatic Impact Assessment Program*, A. J. Broderick, ed., (DOT-TSC-OST-73-4), U.S. Department of Transportation, Washington, D.C., 321-337.

Matsuno, I., 1980, Lagrangian motion of air parcels in the stratosphere and their implications for the general circulation of the stratosphere, *Pure Appl. Geophys.* 118:189-216.

Reed, R. J., and E. F. Danielsen, 1959, Fronts in the vicinity of the tropopause, *Arch. Meteor. Geophys. Bioklam.* A11:1-17.

Reed, R. J., and K. E. German, 1965, A contribution to the problem of stratospheric diffusion by large-scale mixing, *Mon. Weather Rev.* 93:313-321.

Telegadas, K., and R. J. List, 1964, Global history of the 1958 nuclear debris, *J. Geophys. Res.* 69:4741-4753.

Chapter 4

Ozone in the Troposphere

Jack Fishman
NASA-Langley Research Center

HISTORICAL PERSPECTIVE

Although ozone has been known to be a photochemically active species in the stratosphere, the first studies that investigated its distribution and budget within the troposphere assumed that it was a chemically inert species in this region of the atmosphere. Thus, the first tropospheric studies inferred the magnitude of the flux of ozone between the stratosphere and the troposphere by equating it with the amount of ozone destroyed at the earth's surface. It was not until the early 1970s that this viewpoint was challenged, when it was suggested that photochemical processes might play an important role in the tropospheric ozone cycle. At that time a considerable degree of misunderstanding evolved between those who supported the classical viewpoint, which assumed that tropospheric ozone was totally inert, and the school that stated that photochemical processes dominated the tropospheric distribution. As more atmospheric measurements of various trace gases and better chemical kinetic information have become available, and as more realistic photochemical and transport models of the troposphere have been developed, it is now generally agreed that both transport and chemical processes are important in the determination of the tropospheric ozone budget and its observed distribution. The precise interaction among the physical and chemical processes that control the observed tropospheric distribution and that determine the magnitude of the various source and sink terms of this trace gas in the troposphere still is not completely understood, but much progress has been made during the 1970s.

MEASUREMENTS OF TROPOSPHERIC OZONE

Surface Observations

The data base for tropospheric ozone is fairly extensive, but unfortunately most of the measurements have been made at the surface, which may not allow for an accurate assessment of the global distribution of tropospheric ozone. The two most comprehensive studies that have examined the behavior of tropospheric ozone in remote regions of the world have been carried out by the Max Planck Institute for Aeronomy in West Germany (Fabian and Pruchniewicz, 1977) and through the Geophysical Monitoring for Climatic Change Program (GMCC) by the United States National Oceanic and Atmospheric Administration (NOAA) (Oltmans, 1981). Both these studies employed long-term surface-based continuously monitoring ozone instruments to infer a latitudinal distribution of tropospheric ozone. The primary shortcoming of such a monitoring network is the necessary assumption that ozone concentrations within the planetary boundary layer accurately reflect the free tropospheric O_3 concentrations.

The Max Planck network stretched between 70°N (Tromsö, Norway) and 34°S (Hermanus, South Africa) near 20°E longitude. Most of these stations operated on a regular basis for at least 5 years from 1970-1975. From their data, Fabian and Pruchniewicz (1977) inferred free tropospheric ozone concentrations by examining the daily maximum ozone values that were above the mean of all daily maxima for a particular month. This treatment of the data supposedly insured that only data that had undergone sufficient vertical mixing between the surface boundary layer and the free troposphere would be included in the analysis. Such an a priori assumption about the measurements dictates that photochemical generation of ozone within the boundary is negligible. Thus, the analysis, interpretation, and implication of this data set must be viewed with some caution. Some of the conclusions based on this study will be highlighted later in this chapter when the tropospheric ozone budget is discussed in more detail.

The GMCC Program at NOAA has continuously monitored surface ozone for several years at five selected locations: Barrow, Alaska (71°N);

Fritzpeak, Colorado (40°N); Mauna Loa, Hawaii (19.5°N); Samoa (14°S), and the South Pole (90°S). All of these stations are near 160°W longitude. A summary of the annual cycle of surface ozone concentrations is shown in Figure 4-1. As can be seen from this figure, each of the five stations has a different annual cycle that may be linked to its own particular orography and geography and the large-scale dynamic processes that determine the origin of the air masses being sampled. The absolute concentration of

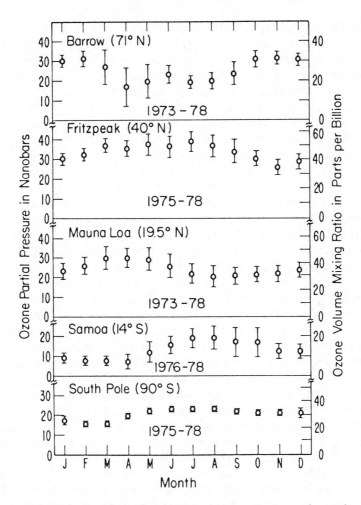

Figure 4-1. Averaged monthly surface ozone amounts. Bars are plus and minus one standard deviation and represent variability within the month. *(From Oltmans, S. J., 1981, Surface ozone measurements in clean air, J. Geophys. Res.* **86;** *copyright © 1981 by the American Geophysical Union.)*

each of these stations varies considerably; highest concentrations are found in the Northern Hemisphere midlatitudes and lowest in the Southern Hemisphere tropics. This general latitudinal pattern has been also noted from aircraft measurements in the free troposphere.

Free Tropospheric Measurements from Ozonesondes

Quality of Tropospheric Ozonesonde Data.   Until recently, very little effort has gone into deriving a distribution of ozone in the free troposphere. The first extensive study that concentrated on the tropospheric ozone distribution was done by Chatfield and Harrison (1977a, 1977b). In their study, they summarized measurements from more than 700 ozonesondes at six stations located near 75°W between 9°N and 53°N. Southern hemispheric ozonesonde data are considerably less abundant. Outside of the ozone measurement program at Aspendale, Australia (Pittock, 1977), which has launched more than 1000 ozonesondes, Fishman, Solomon, and Crutzen (1979) summarize less than 200 soundings over a 15-year period for the rest of the Southern Hemisphere.

There are some serious drawbacks in inferring a global tropospheric distribution from ozonesonde data. First of all, Chatfield and Harrison (1977a) have discussed the inherent differences between the two types of sondes used to measure ozone. Their study suggests that concentrations measured using chemiluminescent ozonesondes may be systematically lower (by as much as 50%) in the troposphere than the values derived from the currently used electrochemical ozonesondes. This is an important point, since the largest single set of ozone data is from the North American Ozonesonde Network (Hering and Borden, 1964-1967) that was in operation between 1963 and 1969. The chemiluminescent sondes were used between 1963 and 1966. Many chemiluminescent sondes showed a clear loss of sensitivity in flight; in others, no loss was evident. When a loss was suspected, a correction factor corresponding to a percentage decrease in sensitivity that was linear in time was applied to the stratospheric portion of the soundings. In addition, high apparent ozone in the troposphere was often associated with those soundings, showing a serious loss of sensitivity in the stratosphere. According to Chatfield and Harrison (1977a) the method and amount of correction to an individual ozonesonde was chosen somewhat arbitrarily. Their primary contention is that if a sonde was known to lose sensitivity during flight, and if the tropospheric values appeared "reasonable," no correction was applied in the troposphere, even though the tropospheric portion of the flight may have been in error.

Such an inconsistency would yield statistically lower values for the chemiluminescent ozonesondes than the tropospheric data derived from electrochemical ozonesondes. On the other hand, it should be noted that Fishman, Solomon, and Crutzen (1979) did not find such a large discrepancy between the two types of sondes used at Panama and Canton Island during this time period although this tropical data base is considerably smaller than the one examined by Chatfield and Harrison (1977a).

Distribution of Tropospheric Ozone Derived from Ozonesonde Data. Despite the limitations of the ozonesonde data base, it is, as stated previously, the most abundant data source for ozone in the free troposphere. In both hemispheres, the greatest amount of data is at midlatitudes. Figure 4-2 depicts the monthly variation of ozone at Boulder (40°N), the combined data from Wallops Island (38°N) and Bedford (42°N), and at Aspendale (38°S). At 200 mb, a distinct annual maximum is present in the late winter and early spring (months 2, 3, 4) at all stations. In addition, the amplitude of the annual cycle and the absolute concentrations of all three data sets are very similar. This feature is a result of downward transport from the stratosphere combined with the fact that the average height of the tropopause is lower during this time of the year (Fishman, Solomon, and Crutzen, 1979).

The seasonal variation of these data sets at 800 mb is not nearly as similar as it was for the upper troposphere. Whereas both Northern Hemisphere data sets vary by nearly a factor of two between late summer and early winter, a seasonal cycle for Aspendale at 800 mb is virtually nonexistent. Furthermore, the average concentration at Aspendale at 800 mb is considerably less than the characteristic concentrations measured in the Northern Hemisphere midlatitudes. This hemispheric difference has been borne out by subsequent aircraft measurements in the middle of the free troposphere.

The ozonesonde measurements in the tropics are considerably fewer than those taken at midlatitudes. Fishman, Solomon, and Crutzen (1979) have summarized 43 soundings from Panama (9°N) and 42 soundings from the Southern Hemispheric tropical locations of Canton Island (2°S) and La Paz (16°S). Figure 4-3 shows the average vertical distribution of tropospheric ozone in the Northern and Southern Hemisphere tropical regions. The observed difference, which is on the order of 50%, is consistent with the surface measurements discussed in the previous section and has likewise been substantiated by measurements aboard airplanes and ships, which indicate that the minimum in the latitudinal distribution of tropospheric ozone is located in the tropical region or the Southern Hemisphere.

Aircraft Measurements of Tropospheric Ozone

The results from a measurement program that had as one of its objectives an investigation of the latitudinal distribution of tropospheric ozone were not reported until Routhier et al. (1980) summarized the ozone data obtained during the 1977 and 1978 flights of Project GAMETAG (Global

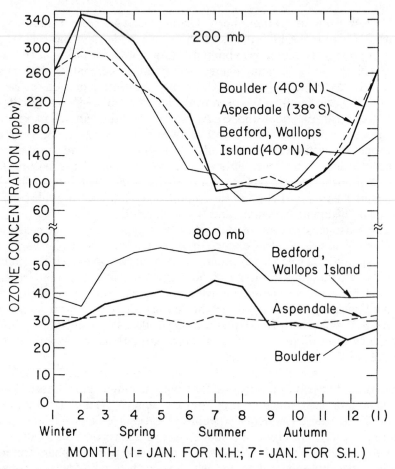

Figure 4-2. Monthly variation of ozone for three data sets in the midlatitudes. The upper curves depict the variation in the upper troposphere (200 mb), whereas the lower set of curves describes the seasonal variations in the lower free troposphere at 500 mb *(From Fishman, J., S. Solomon, and P. J. Crutzen, 1979, Observational and theoretical evidence in support of a significant in-situ photochemical source of tropospheric ozone, Tellus 32.)*

Atmospheric Measurement Experiment of Tropospheric Aerosols and Gases). Subsequently, Seiler and Fishman (1981) have reported a tropospheric O_3 data set using measurements obtained during a 6-week series of flights in July and August, 1974. Although the original intent of the 1974 measurements was to examine the vertical and latitudinal distributions of CO, H_2, and CO_2, a carefully calibrated O_3 instrument was

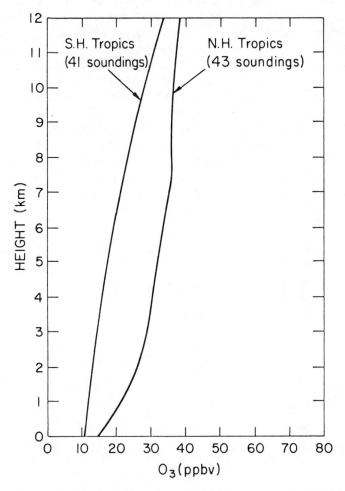

Figure 4-3. Average vertical profiles of tropospheric ozone in the Northern and Southern Hemisphere tropical regions. *(From Fishman, J., S. Solomon, and P. J. Crutzen, 1979, Observational and theoretical evidence in support of a significant in-situ photochemical source of tropospheric ozone, Tellus **32.**)*

aboard so that it could be readily determined whether or not the air being sampled was in the stratosphere or the troposphere. Whereas the GAMETAG data set consists primarily of measurements at only one flight altitude (~5.5 km), the Seiler and Fishman (1981) analysis was derived from more than 70 vertical profiles between flight level of a jet airplane (8-12 km) and the surface with an average latitudinal resolution of about 8°. Between the two field programs, data were obtained between 57°S and 67°N.

The two-dimensional distribution of tropospheric ozone during August 1974, from the Seiler and Fishman (1981) study, is shown in Figure 4-4. This figure confirms the hemispherically asymmetrical distribution of tropospheric ozone that was first illustrated in the study of Fishman and Crutzen (1978), but that had been based on a less reliable data set derived from ozonesonde measurements. Figure 4-5 compares the tropospheric O_3 data sets of Fishman and Crutzen (1978), Routhier et al. (1980) and Seiler and Fishman (1981). For a fair comparison, one should note that the open and closed circles (from Seiler and Fishman [1981] and Fishman and Crutzen [1978] respectively) represent average tropospheric column concentrations, whereas the triangles and crosses (from Seiler and Fishman [1981] and Routhier et al. [1980] respectively) represent the observed O_3 concentration at ~6 km altitude. One feature common to each of these

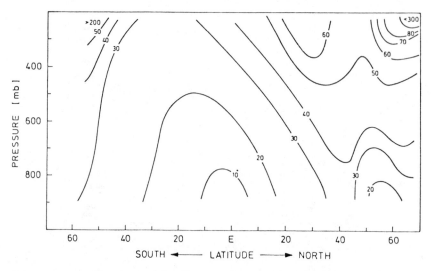

Figure 4-4. Two-dimensional distribution of ozone obtained during August, 1974. *(From Seiler, W., and J. Fishman, 1981, The distribution of carbon monoxide and ozone in the free troposphere, J. Geophys. Res. **86**; copyright © 1981 by the American Geophysical Union.)*

latitudinal profiles is the minimum found near 10°S. In addition, each profile suggests that 30-50% more ozone is present in the Northern Hemisphere than in the Southern Hemisphere. Some of the differences seen between the individual data sets most likely reflect the inherent variability that tropospheric ozone exhibits.

One feature that is present in many tropospheric ozone profiles is the degree of layering that exists (Routhier et al., 1980) but that seems to be present almost exclusively in the Northern Hemisphere (Seiler and Fishman, 1981). In the 1977 and 1978 GAMETAG flights, tropospheric O_3 profiles were obtained concurrently with a fast-response hygrometer. An example of the types of vertical profiles that were obtained using these instruments is shown in Figure 4-6. It is particularly noteworthy that these profiles exhibit a significant anticorrelation between O_3 concentration and dew point temperature. A similar O_3 layering has been

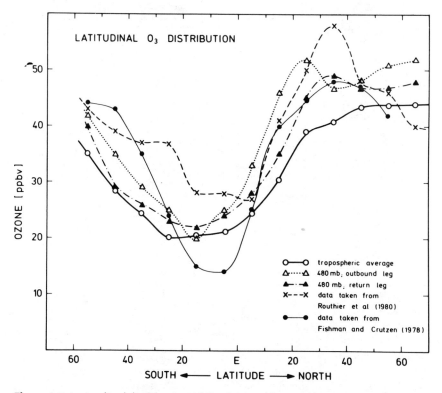

Figure 4-5. Latitudinal distribution of O_3 obtained from Seiler and Fishman (1981), denoted by open circles, and compared with measurements from other studies.

reported by Fishman, Seiler, and Haagenson (1980) as depicted in Figure 4-7. Carbon monoxide (CO) was measured simultaneously with a fast-response sensor, and its vertical structure is likewise shown in this figure. Of particular interest in this figure is the concurrent presence of elevated concentrations of O_3 and CO in the free troposphere at altitudes of 3.8 and 7.6 km. Fishman and Seiler (1983) analyzed 70 pairs of simultaneous CO and O_3 profiles and noted that most of the fluctuations of these trace gases in the Northern Hemisphere were positively correlated. In the Southern Hemisphere, the degree of vertical variability was considerably less pronounced (e.g., see Fig. 4-8), and no strong positive or negative correlation was found to be present for O_3 and CO.

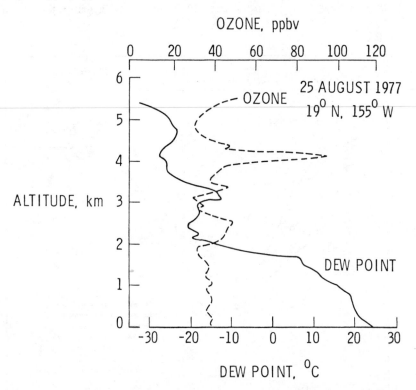

Figure 4-6. Representative O_3 and H_2O layering in the subtropical-tropical Pacific Ocean. *(After Routhier, F., and D. D. Davis, 1980, Free tropospheric/boundary-layer airborne measurements of H_2O over the latitude range of 58°S to 70°N: Comparison with simultaneous ozone and carbon monoxide measurements, J. Geophys. Res.* **85.***)*

OZONE PHOTOCHEMISTRY IN THE
FREE TROPOSPHERE

Ultraviolet radiation in the 300-320 nm wavelength region can make its way into the free troposphere where ozone can be photolyzed to produce atomic oxygen in the metastable $O(^1D)$ state. Most of the metastable oxygen will return to the ground state $O(^3P)$ via a collision with a neutral molecule; in turn, nearly all $O(^3P)$ atoms will recombine with molecular oxygen to yield O_3 again. Thus, the tropospheric analogy of "odd-oxygen" chemistry found in the stratosphere is of little importance and results in either of the two "do-nothing" cycles: first,

$$O_3 + h\nu \rightarrow O(^1D) + O_2, \lambda < 320 \text{ nm},$$

followed by

$$O(^1D) + M \rightarrow O(^3P) + M,$$

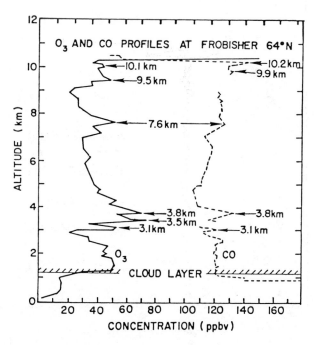

Figure 4-7. Simultaneous profiles of ozone and carbon monoxide. *(From Fishman, J., W. Seiler, and P. Haagenson, 1980, Simultaneous presence of O_3 and CO bands in the troposphere, Tellus **32**.)*

and terminated by

$$O(^3P) + O_2 + M \rightarrow O_3 + M;$$

and second,

$$O_3 + h\nu \rightarrow O(^3P) + O_2, \qquad \lambda < 1140 \text{ nm}$$

terminated by

$$O(^3P) + O_2 + M \rightarrow O_3 + M.$$

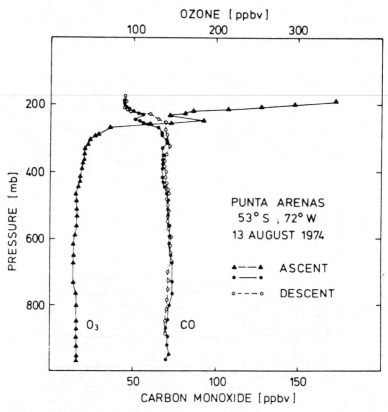

Figure 4-8. Vertical profiles of O_3 and CO constructed off the southern coast of South America. *(From Seiler, W., and J. Fishman, 1981, The distribution of carbon monoxide and ozone in the free troposphere, J. Geophys. Res. 86; copyright © 1981 by the American Geophysical Union.)*

The odd-oxygen chemistry is unimportant in the troposphere for two reasons. First, molecular oxygen O_2 cannot be photodissociated in the troposphere since radiation below 242 nm is required and no such high energy photons can reach the troposphere. Therefore, atomic oxygen concentrations (both in the singlet-D and triplet-P states) do not become very large. Second, because of the much greater abundance of both N_2 and O_2 (the most abundant neutral molecules) in the troposphere, the recombination of $O(^3P)$ to give back O_3 is much more efficient.

However, this nonproductive odd-oxygen sequence can be broken if $O(^1D)$ reacts with a non-neutral molecule such as H_2O. In this case,

$$O_3 + h\nu \rightarrow O(^1D) + O_2, \qquad \lambda < 320 \text{ nm}$$

followed by

$$\frac{O(^1D) + H_2O \rightarrow 2\,OH}{O_3 + H_2O \rightarrow 2\,OH + O_2} \quad \text{(net reaction sequence)}$$

In this series of reactions, two OH radicals have been produced, and a photochemical sink for tropospheric odd-oxygen (ozone) has been established. Furthermore, OH can react with O_3 to set up the catalytic destruction sequence

$$OH + O_3 \rightarrow HO_2 + O_2$$

and

$$\frac{HO_2 + O_3 \rightarrow OH + 2\,O_2}{2\,O_3 \rightarrow 3\,O_2} \quad \text{(net reaction sequence)}.$$

Using a one-dimensional model of the troposphere, Fishman and Crutzen (1977) estimated that approximately one-half the ozone entering the troposphere from the stratosphere would be destroyed by the above photochemical reactions before reaching the ground. However, once the OH is formed, it is probable that it will react with either CO or CH_4 rather than with O_3 in the troposphere (e.g., see Logan et al., 1981). In fact, reaction with OH is the primary removal mechanism of CO, CH_4, and many other trace gases found in the troposphere. It is the oxidation of these tropospheric trace gases by OH that can eventually lead to in-situ photochemical generation of tropospheric ozone.

The two most discussed reaction sequences that may lead to photochemical O_3 generation in the free troposphere are the oxidation of CO via

$$CO + OH \rightarrow CO_2 + H$$

$$H + O_2 + M \rightarrow HO_2 + M$$

$$HO_2 + NO \rightarrow NO_2 + OH$$

$$NO_2 + h\nu \rightarrow NO + O, \qquad \lambda < 420 \text{ nm}$$

$$\underline{O + O_2 + M \rightarrow O_3 + M}$$

$$CO + 2\,O_2 \rightarrow CO_2 + O_3 \quad \text{(net reaction sequence)},$$

and the oxidation of CH_4 through the sequence

$$CH_4 + OH \rightarrow CH_3 + H_2O$$

$$CH_3 + O_2 + M \rightarrow CH_3O_2 + M$$

$$CH_3O_2 + NO \rightarrow CH_3O + NO_2$$

$$CH_3O + O_2 \rightarrow CH_2O + HO_2$$

$$HO_2 + NO \rightarrow OH + NO_2$$

$$2 \times (NO_2 + h\nu \rightarrow NO + O, \qquad \lambda < 420 \text{ nm})$$

$$\underline{2 \times (O + O_2 + M \rightarrow O_3 + M)}$$

$$CH_4 + 4\,O_2 \rightarrow H_2O + CH_2O + 2\,O_3 \quad \text{(net reaction sequence)}.$$

The key to the ozone-yielding efficiency of both of these oxidation sequences is the availability of NO (nitric oxide). This point will be addressed in considerably more detail in the next section, where the budget of tropospheric O_3 is discussed.

On a more localized scale, in-situ photochemical ozone production can become very large if high concentrations of hydrocarbons and nitric oxide are present. In the Los Angeles area, for example, O_3 concentrations of greater than 500 ppbv have been observed. The "smog" chemistry taking place is somewhat analogous to the chemistry described in the sequences outlined for CO and CH_4. In the case of smog, however, it is the presence of a large amount of nonmethane hydrocarbons that can be oxidized that results in the formation of the high concentrations of O_3. The primary difference is that nonmethane hydrocarbons react much faster with oxidizing species (such as OH) that will yield several peroxy radicals (analogous to HO_2 and CH_3O_2) much more rapidly than HO_2 or CH_3O_2 is produced from CO and CH_4 oxidation. The key to O_3 formation is the conversion of NO to NO_2 by one of these reactions:

Table 4-1. Tropospheric Ozone Budget

Source or Sink	Magnitude, 10^{11} mol cm^{-2}s^{-1}		References
	Northern Hemisphere	Southern Hemisphere	
Transport from the stratosphere	0.5-0.8	0.3-0.4	a, b, c
Destruction at the ground	0.7-1.4	0.4-0.8	c, d
Photochemical destruction	1.9-2.2	1.1-1.2	e, f, g, h
Photochemical production	1.9-3.1	1.1-2.9	e, f, g, h

[a] Danielsen and Mohnen (1977)
[b] Mahlman, Levy, and Maxim (1980)
[c] Gidel and Shapiro (1980)
[d] Galbally and Roy (1980)
[e] Fishman, Solomon, and Crutzen (1979)
[f] Liu et al. (1980)
[g] Logan et al. (1981)
[h] Chameides and Tan (1981)

$$RO_2 + NO \rightarrow NO_2 + RO$$
$$NO_2 + h\nu \rightarrow NO + O, \qquad \lambda < 420 \text{ nm}$$
$$\underline{O + O_2 + M \rightarrow O_3 + M}$$
$$RO_2 + O_2 \rightarrow RO + O_3$$

where RO_2 and RO represent higher order radicals. In summary, if NO can be converted to NO_2 by any species other than O_3 itself, the net result will be a photochemical production of ozone in the troposphere. This process is true in both the free troposphere and the polluted urban boundary layer.

THE TROPOSPHERIC OZONE BUDGET

The components of the global tropospheric ozone budget can be divided into four general categories: transport from the stratosphere, destruction at the earth's surface, in-situ photochemical destruction, and in-situ photochemical production. Table 4-1 summarizes some of the previous estimates of the magnitude of each of these sources and sinks. As can be seen from this table, each of the four components of the budget is fairly comparable in magnitude. A detailed examination of the quantification and the uncertainty inherent in the calculation of each of these terms will be presented in the remainder of this section.

Ozone Destruction at the Earth's Surface

The destruction of ozone at the earth's surface plays an important role in the budget of tropospheric ozone. (In addition, it is the only quantity in the budget that can be directly measured.) The earlier estimates of surface destruction were based on the "box method," in which ozone was placed in a box with one side open against the ground and the time rate of change of the O_3 concentration was measured (e.g., Aldaz, 1969; Galbally and Roy, 1980). From these measurements, representative deposition velocities were derived for certain characteristic surfaces.

One way of interpreting the results of these early deposition velocity experiments is to draw an analogy between the reciprocal of the deposition velocity and a series of resistances, for the transfer of O_3 to the surface (see Fig. 4-9). In this figure, the aerodynamic resistance r_a is a measure of the efficiency of the atmosphere in transporting ozone to the ground where it can interact with surface elements. The surface resistance r_s is a measure of the ability of O_3 to contact the surface from a small distance away from the surface and is most likely a function of the molecular diffusivity of ozone. Lastly, the canopy stomatal resistance r_c is most characteristic of the type of vegetation and is a measure of the efficiency of the plant material's ability to react with ozone.

Extrapolation of these box measurements for use in global-scale calculations of ozone deposition necessitates two important cautions. Fabian

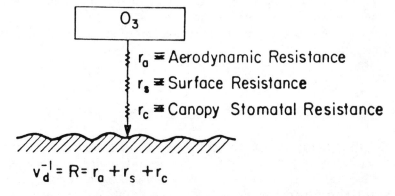

Figure 4-9. An electrical analog for visualizing the concept of a deposition velocity as the reciprocal of a resistance. This resistance can then be further divided into the sum of three series resistances. *(From Lenschow, D. H., and E. W. Barry, 1980, Photochemical modeling: Aircraft data, The CHON Photochemistry of the Troposphere, Notes from a Colloquium, A. C. Delaney, ed. [NCAR/CQ-7 + 1980-ASP, available from the National Center for Atmospheric Research, Boulder, Colo.].)*

and Junge (1970) noted that an artificial stirring of the air within the box will create a loss rate within the box that is higher than what should exist in the natural environment. In the Aldaz (1969) experiments, the air inside the box is perfectly stirred to make sure that the ozone density remains uniform within the volume of the box. In the real atmosphere, however, a sharp vertical gradient of ozone is present in the lowest meter, and therefore only a fraction of the average concentration within the box actually comes in contact with the ground. With respect to the diagram shown in Figure 4-9, Fabian and Junge (1970) contend that the Aldaz (1969) experiments did not realistically account for the aerodynamic resistance, r_a and therefore that Aldaz's experimental results must be modified before use in a global calculation. Secondly, Galbally and Roy (1980) have pointed out that ozone uptake by vegetation is highly dependent upon the stomatal resistance of a particular plant. In general, stomata are closed during nighttime, which results in a higher stomatal resistance r_c and a lower equivalent deposition velocity during the night than during the day.

More recently, however, instrumentation has been developed to measure the flux of ozone directly using an eddy correlation technique. To achieve these measurements, an ozone sensor must be used that combines both fast response (10 Hz) and high enough sensitivity to detect small fluctuations (Pearson and Stedman, 1980). In conjunction with the use of a fast response vertical velocity sensor, the vertical flux of ozone can be measured directly by measuring the covariance of the fluctuations of these two quantities; thus

$$\text{flux of } O_3 = \overline{w'[O_3]'},$$

where w' is the fluctuation of the vertical velocity from the mean value of the vertical velocity w, and $[O_3]'$ is a similar term for the concentration of ozone. The overbar bar represents the average value of the product of these fluctuations over the sampling time. In practice, the O_3 and w sensors aboard an airplane should resolve fluctuations on the order of 0.1 s, and a typical sampling time is about 15 min. From the covariance measurements, a deposition velocity w_d can be computed from

$$w_d = \frac{\overline{w'[O_3]'}}{\overline{[O_3]}},$$

where $\overline{[O_3]}$ is the average concentration of ozone during the sampling period. This flux measurement procedure, commonly referred to as the eddy correlation method, has the advantage that the natural surround-

Table 4-2. Loss Rate of Ozone at the Surface

Surface	Deposition Rate	Method	References
Snow	0.02 cm s^{-1}	Box	Aldaz (1969)
	0.10	Box	Galbally and Roy (1980)
	0.03	Eddy correlation (tower)	Wesely, Cook, and Williams (1981)
Oceans	0.04	Box	Aldaz (1969)
	0.10	Box	Galbally and Roy (1980)
	0.06	Eddy correlation (airplane)	Lenschow, Pearson, and Stankov (1982)
Grassland	0.60	Box	Aldaz (1969)
	0.63 (day)	Box	Galbally and Roy (1980)
	0.25 (night)	Box	Galbally and Roy (1980)
	0.47	Eddy correlation (airplane)	Lenschow, Pearson, and Stankov (1981)
	0.67	Eddy correlation (airplane)	Pearson et al. (1982)
Forest	0.87 (day)	Box	Galbally and Roy (1980)
	0.26 (night)	Box	Galbally and Roy (1980)
	2.0	Eddy correlation (airplane)	Lenschow, Pearson, and Stankov (1982)

ings in which the measurements are being made are not disturbed. Thus, accurate ozone flux measurements can be made from an instrumented tower or from an airplane flying within the planetary boundary layer during those portions of the day when the boundary layer is well mixed and, therefore, the flux is constant throughout.

A summary of some of the studies that have measured deposition velocities onto various types of surfaces is presented in Table 4-2. This list is not meant to be complete, but it is representative of the values derived by various authors using different techniques. It should also be noted that average values for the studies referenced in this table are listed but that much variability exists within a particular experiment. For a more comprehensive analysis of the data and a more thorough review of previous studies the reader is referred to Wesely, Cook, and Williams (1981) and Galbally and Roy (1980).

By using the methodology described in Galbally and Roy (1980), the loss rate of O_3 at the earth's surface is computed and presented in Table 4-3. To calculate the surface removal rate, the following average deposition velocities have been utilized: 0.06 cm s^{-1} for oceans and snow; 0.4 cm s^{-1} for grasslands; and 0.8 cm s^{-1} for forests. Critical in this calculation is the latitudinal distribution of ozone, which has been obtained using the five data sets referenced in Table 4-3.

Table 4-3. Ozone Destruction at the Earth's Surface Computed by Methodology Described in Galbally and Roy (1980)

Latitude Belt and Surface	% Global Area		O_3 Concentration, ppbv (references)	Loss Rate
90-60°S				
Snow—Ocean	7		20 (a)	$3.0 \times 10^{10} \, \text{mol cm}^{-2} \text{s}^{-1}$
60-30°S				
Ocean	17			
Grassland	0.9	18	24 (b, c, d)	4.1
Forest	0.1			
30-15°S	'			
Ocean	,9			
Grassland	2.5	12	23 (c, d)	9.3
Forest	0.5			
15-0°S				
Ocean	10			
Grassland	1.4	13	12 (a, c, d)	5.6
Forest	1.6			

Southern Hemispheric Average: $5.6 \times 10^{10} \, \text{mol cm}^{-2} \, \text{s}^{-1}$

0-15°N				
Ocean	10			
Grassland	1.8	13	19 (c, d, e)	8.6
Forest	1.2			
15-30°N				
Ocean	8			
Grassland	3.8	12	26 (a, c, d, e)	11.7
Forest	0.2			
30-60°N				
Ocean	9			
Grassland	5.8	18	34 (a, c, d, e)	25.5
Forest	3.2			
60-90°N				
Ocean—Snow	3.8			
Grassland	2.2	7	28 (a, c, d, e)	18.9
Forest	1.2			

Northern Hemispheric Average: $16.9 \times 10^{10} \, \text{mol cm}^{-2} \, \text{s}^{-1}$

Global Average: $11.2 \times 10^{10} \, \text{mol cm}^{-2} \, \text{s}^{-1}$

[a] Oltmans (1981)
[b] Pittock (1974)
[c] Seiler and Fishman (1981)
[d] Routhier et al. (1980)
[e] Chatfield and Harrison (1977b)

An important result of the surface loss rates summarized in this table is the relatively high value calculated for the Northern Hemisphere (refer to Table 4-1 for a summary of previous estimates). The primary reason for this is the use of somewhat higher concentrations in the Northern Hemisphere than had been previously utilized. For example, the calculations presented in this table use values of 19, 26, 34, and 28 ppbv for the surface concentrations in the four Northern Hemisphere boxes, whereas Galbally and Roy (1980) used values of 12, 17, 19 and 21 ppbv. As a result, the Northern Hemisphere surface concentrations used in Table 4-3 are 61% higher than those used by Galbally and Roy and consequently the destruction rate in the Northern Hemisphere is 67% greater in these calculations.

Another noteworthy point of Table 4-3 is the large asymmetry between the loss rates in the two hemispheres. This ratio of 3.0 (Northern to Southern Hemisphere) is larger than ratios calculated in most previous studies and is primarily a result of the higher Northern Hemisphere loss rate summarized in Table 4-3. The loss rate for the Southern Hemisphere of 5.6×10^{10} mol cm^{-2} s^{-1} is slightly lower than the value of 6.9×10^{10} mol cm^{-2} s^{-1} derived by Galbally and Roy (1980) and is primarily attributable to the lower deposition velocity used for oceans in this study (0.06 cm s^{-1}). The Galbally and Roy study used 0.1 cm s^{-1} for the deposition velocity over water. Using the same deposition rates as the ones summarized by Galbally and Roy and the observed O$_3$ distribution summarized in Table 4-3 of this study produces a Northern to Southern Hemisphere asymmetry of 2.4 (17.4×10^{10} vs. 7.2×10^{10} mol. cm^{-2} s^{-1}). Thus, even though the more recent global estimates of the amount of loss at the earth's surface seem to be coming close to a value of 10×10^{10} mol cm^{-2} s^{-1} ($\pm 30\%$), the global distribution of the sink is still not well established. An understanding of the distribution of the loss at the earth's surface is critical if an accurate picture of the tropospheric ozone budget is ever to be obtained, and further measurements are necessary if the uncertainty of this distribution is to be minimized.

Stratospheric Source of Tropospheric Ozone

As mentioned previously, the transport of ozone from the stratosphere into the troposphere was at one time considered to be the only source of tropospheric ozone: it was assumed that the chemical reactivity of tropospheric O$_3$ in all regions other than a polluted environment was so small in the troposphere that it could be excluded from the global tropospheric O$_3$ cycle. Thus, the early studies that examined stratosphere-troposphere exchange did so by calculating the rate of destruction of O$_3$ at the ground with the assumption that the quantity destroyed at the ground was equivalent to what had been transported across the tropopause (e.g., Fabian, 1973; Fabian and Pruchniewicz, 1977). A direct measurement of the amount of O_3 entering the troposphere from the stratosphere is considerably more difficult.

Danielsen (1964) has summarized a detailed examination of mass and ozone transfer processes between the stratosphere and the troposphere. During the early 1960s, radioactive tracers (results of atmospheric atomic bomb testing) could be used to estimate mass outflow from the stratosphere into the troposphere. Thus, in the 1960s a knowledge of the surface deposition of ^{90}Sr in conjunction with an aircraft sampling program that measured the distribution of ^{90}Sr and ozone in the lower stratosphere and upper troposphere could be used to estimate the amount of ozone in the troposphere that must have had its origin in the stratosphere. Such an analysis yielded a value of 7.8×10^{10} mol cm^{-2} s^{-1} as a Northern Hemisphere average, and it was also deduced that this outflow rate maximizes in mid-April where the outflow rate of ozone is approximately five times larger than the amount coming from the stratosphere in late autumn (Danielsen and Mohnen, 1977).

The studies of Viezee, Johnson, and Singh (1979) and Mohnen and Reiter (1977) analyzed a number of cases in which ozone entered the troposphere through the stratosphere and estimated the mass of stratospheric air that entered the tropopause per cyclogenic event. By assuming an average O_3 mixing ratio and by estimating the number of such events taking place on a yearly basis, they obtained comparable results of 8×10^{10} mol cm^{-2} s^{-1}. In another case study, Shapiro (1980) measured the ozone flux directly on an airplane flying above and below the tropopause using the eddy correlation method discussed in the previous section. In his analysis of this cyclogenesis event, Shapiro pointed out that some of the ozone transported downward on the anticyclonic side of the jet stream reentered the stratosphere on the cyclonic side of the jet. Shapiro's findings suggest that turbulent mixing processes in the vicinity of the jet stream should be considered when quantifying the amount of stratospheric ozone coming into the troposphere. Clearly, there are not enough measurements to quantify the amount of ozone transported from the stratosphere to the troposphere, but an extrapolation of the few direct measurements supports a value of $4\text{-}8 \times 10^{10}$ mol cm^{-2} s^{-1} in the Northern Hemisphere. No similar calculations have been made for the Southern Hemisphere.

Another method of estimating the ozone flux across the tropopause is through the use of sophisticated numerical general circulation models (GCMs) of the atmosphere. Mahlman et al. (1980) have reported the results of one such study using the Geophysical Fluid Dynamics Laboratory GCM, which looked at the cross-tropopause transport of a stratified "ozone-like" tracer. In their numerical experiment, they compute a cross-tropopause flux of 6.6×10^{10} and 3.6×10^{10} mol cm^{-2} s^{-1} for the Northern and Southern Hemispheres respectively. In another study, Gidel and Shapiro (1980) utilized the National Center for Atmospheric Research GCM to examine stratosphere-troposphere exchange. In their experiment, they actually calculated the distribution and cross-tropopause flux of potential vorticity, which should be proportionally similar to the distribution and subsequent transport of ozone (cf. Danielsen, 1968). Both O_3 mixing ratios and potential vorticity are very nearly conserved in the lower stratosphere, and each has a very strong vertical stratification in this region of the atmosphere. Using this technique, Gidel and Shapiro estimated a cross tropopause flux of ozone of 4.9×10^{10} mol cm^{-2} s^{-1} in the Northern Hemisphere and 2.5×10^{10} mol cm^{-2} s^{-1} in the Southern Hemisphere.

Although the methods of estimating cross-tropopause flux are quite different, it is interesting that each study has arrived at a Northern Hemisphere flux value of $6 \pm 2 \times 10^{10}$ mol cm^{-2} s^{-1}. In addition, both model studies indicate considerably more cross-tropopause flux ($\sim 90\%$) in the Northern Hemisphere than in the Southern Hemisphere. If one accepts the estimated hemispheric destruction rates listed in Table 4-3, there is considerably more destruction at the surface in the Northern Hemisphere than is transported from the stratosphere, implying that an additional mechanism that yields a net production of $11 \pm 8 \times 10^{10}$ mol cm^{-1} s^{-1} is necessary to balance the tropospheric O_3 budget in the Northern Hemisphere. In the Southern Hemisphere, the transport from the stratosphere and the destruction rate at the ground appear to be more in balance. To balance the budget in this hemisphere, an additional source of $3 \pm 5 \times 10^{10}$ mol cm^{-2} s^{-1} is required; part of this source would be transport from the Northern Hemisphere.

Photochemical Destruction of Tropospheric Ozone

The photochemical destruction of ozone takes place primarily through the two sequences of reactions

$$O_3 + h\nu \rightarrow O(^1D) + O_2, \qquad \lambda < 320 \text{ mn}$$

followed by

$$\frac{O(^1D) + H_2O \rightarrow 2\ OH,}{O_3 + H_2O \rightarrow O_2 + 2\ OH} \quad \text{(net reaction sequence)},$$

and

$$O_3 + OH \rightarrow HO_2 + O_2$$

followed by

$$\frac{HO_2 + O_3 \rightarrow OH + 2\ O_2}{2\ O_3 \rightarrow 3\ O_2 \quad \text{(net reaction sequence)}.}$$

The direct measurement of O_3 destruction through these processes is quite difficult, since the rate of destruction is only a few ppbv per day. In a measurement experiment, this loss rate could not be distinguished from the day-to-day variability of O_3 due to meteorological influences. Thus, the amount of O_3 destroyed photochemically must be calculated with the use of theoretical photochemical models.

The studies summarized in Table 4-1 that have computed the loss rate of tropospheric O_3 have used two-dimensional diagnostic models to calculate this quantity. In these calculations the long-lived trace gases (i.e., H_2O, CO, O_3, $NO + NO_2$, CH_4, H_2) are prescribed as a function of altitude and latitude and as a function of month or season if an annual cycle is known to be present. In addition, the incoming solar flux as a function of wavelength must also be known, and the effects of absorption and scattering of this radiation must also be properly parameterized. Of all the above inputs that go into a diagnostic photochemical model of tropospheric chemistry, the distribution of NO_x ($NO_x = NO + NO_2$) is probably least understood. Fortunately, the calculation of the tropospheric loss rate of tropospheric O_3 is not overly dependent upon a knowledge of the distribution of NO_x, but on the other hand the rate of in-situ photochemical production of tropospheric ozone is very sensitive to the assumed distribution of NO_x (see the discussion of Fishman, Solomon, and Crutzen, 1979). This latter point will be discussed in detail in the next section.

The primary reactions that comprise the photochemical loss of tropospheric ozone are shown as a function of altitude at two representative Northern Hemisphere latitudes in Figure 4-10 (Logan et al., 1981). As can be seen from this depiction, the most efficient loss mechanism of odd-oxygen is the $O(^1D)$ reaction with water vapor in the lower tropical troposphere. There are two reasons for the large loss rate in this region of the troposphere. First, the amount of H_2O in the marine tropical boundary layer is considerably higher than in any other region of the troposphere, with a volume mixing ratio of more than 2×10^{-2}. The average water content in the troposphere decreases rapidly with altitude and is more than a factor of ten less at an altitude of 5 km. In the temperate latitudes, the water content in the planetary boundary varies considerably during the year, but it is usually less than 1×10^{-2}. Second, more solar radiation is received in the tropics, especially in the 300-320 nm wavelength region where most of the $O(^1D)$ production takes place. This is true both because of a lower average zenith angle in low latitudes and because of the relatively lower overlying stratospheric ozone burden in the tropics. For example, the O_3 photolysis rate to form $O(^1D)$ with a 15° zenith angle and an overlying O_3 burden of 260 Dobson Units (typical of the tropics) is ~6×10^{-5} s^{-1}, whereas the photolysis rate with a 330 D.U. overburden and a 45° zenith angle is only ~2.5×10^{-5} s^{-1} (Dickerson et al., 1979/1980). This difference, coupled with the very large water content in the tropical boundary layer, offsets the fact that less O_3 is present in this region of the atmosphere. At 45°N, however, the reaction of HO_2 and O_3 is the dominant loss mechanism throughout the entire Northern Hemisphere troposphere. Integrated throughout the entire hemisphere it is estimated that $H_2O + O(^1D)$ accounts for 52% of the photochemical loss of tropospheric O_3; $HO_2 + O_3$ for 42%; and $O_3 + OH$ for 6% (Fishman, Solomon, and Crutzen, 1979). The two-dimensional diagnostic studies of Fishman, Solomon, and Crutzen (1979), Liu et al. (1980), Logan et al. (1981) and Chameides and Tan (1981) calculate fairly consistent photochemical loss rates $21 \pm 4 \times 10^{10}$ mol cm^{-2} s^{-1} in the Northern

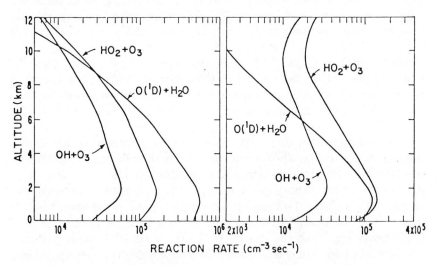

Figure 4-10. Diurnally averaged loss rates of tropospheric odd-oxygen. *(From Logan, J. A., M. J. Prather, S. C. Wofsy, and M. B. McElroy, 1981, Tropospheric chemistry: A global perspective, J. Geophys. Res. **86**; copyright © 1981 by the American Geophysical Union.)*

Hemisphere. Because of the smaller abundance of ozone in the Southern Hemisphere, there is considerably less photochemical destruction, and the diagnostic models calculate a loss rate of $11 \pm 3 \times 10^{10}$ mol cm^{-2} s^{-1}.

Photochemical Production of Tropospheric Ozone

In the polluted urban boundary layer, it is not uncommon to find O_3 mixing ratios in excess of 100 ppbv. The "smog" chemistry that produces the ozone in this environment is a result of large emissions of hydrocarbons and NO_x in the presence of sunlight. When a meteorological condition is present which traps the air within a shallow layer (less than 1 km) near the ground, these pollutants cannot be ventilated and large quantities of photochemically produced ozone can be measured near the surface. In principle, however, the same smog chemistry in the polluted environment should exist in the natural troposphere. In the unpolluted troposphere, the dominant hydrocarbon that is present in methane (CH_4) and it was hypothesized by Crutzen (1973, 1974) that methane oxidation, in the presence of NO_x, should yield a globally integrated production mechanism of tropospheric ozone that is comparable to the amount transported across the tropopause.

Building upon the hypothesis put forth by the work of Crutzen, Chameides and Walker (1973, 1976) published two studies that claimed that the observed distribution of tropospheric ozone could be substantiated by assuming that ozone was in photochemical equilibrium (Chameides and Walker, 1973) and that its diurnal behavior could be explained solely by methane oxidation. Both of these studies were refuted strongly by the supporters of the classical view (Fabian, 1974; Chatfield and Harrison, 1976). Although there were shortcomings in the Chameides and Walker studies, they were forced to make some assumptions that were consistent with the data available at the time of the study but that have now been shown to be incorrect. In particular, they assumed a global NO_x background concentration on the order of 10 ppbv, which was based on the studies of Robinson and Robbins (1969); no knowledge of the vertical distribution of NO_x was available. It would not be until the late 1970s that background levels of NO_x would be found to be 100-1000 times lower than the values put forth by Robinson and Robbins.

On the other hand, there were some aspects of the Chameides and Walker studies that were not consistent with some of the available measurements. For example, they assumed a nitric acid (HNO_3) concentration of 80 ppbv for their results to be valid. No such concentrations had ever been observed, nor has it since, even in the most polluted environments.

In addition, the initial studies of Crutzen never maintained that ozone existed in the troposphere as a species in photochemical equilibrium as assumed by the Chameides and Walker (1973) study.

Another shortcoming of all the earlier photochemical studies of tropospheric ozone (i.e., Crutzen, 1973, 1974; Chameides and Walker, 1973, 1976) was that none of them included any transport mechanism in their calculations. In the next generation of model studies, the one-dimensional photochemical experiments of Chameides and Stedman (1977), Liu (1977), and Fishman and Crutzen (1977) parameterized vertical transport with an eddy diffusion coefficient. Each of these models calculated that in-situ photochemical production of tropospheric O_3 was comparable to the source from the stratosphere. Subsequent to the published results of the above modeling studies, Howard and Evenson (1977) published some new laboratory kinetic data that essentially made all the results of the above theoretical studies null and void. For the first time, the rate of the reaction $HO_2 + NO \rightarrow OH + NO_2$ was measured directly, and it was 15-40 times faster than had been previously believed. The inclusion of this faster rate constant in these modeling studies would have produced too much ozone on the global scale to be consistent with the observations. The only way to reconcile the global tropospheric ozone budget was to hypothesize that considerably less NO_x must be present in the troposphere (Fishman and Crutzen, 1977 ["note added in proof"]; Chameides, 1978). This notion was substantiated by the tropospheric measurements of Noxon (1978) and McFarland et al. (1979) of NO_2 and NO, respectively. Thus, the key to understanding the importance of in-situ photochemical production on the tropospheric ozone budget is the attainment of the distribution of tropospheric NO_x.

The data base for NO and NO_2 is far too sparse to derive a realistic global distribution of these species. However, the wide range of variability found in the measurements suggests that a representative distribution of NO_x may not exist. For example, Bollinger et al. (1982) cite local surface daytime NO_x measurements in Colorado (at an elevation of 3 km) of 0.014 ppbv, whereas Ehhalt and Drummond (1982) suggest that a representative continental boundary layer NO_x mixing ratio should be on the order of 1 ppbv. Thus, even in a fairly well-defined locale, such as "the continental boundary layer of northern mid-latitudes," the spatial and temporal variability of this trace species may not permit the term *representative* to be meaningful. On the other hand, there most likely is considerably less NO_x in the tropics and Southern Hemisphere, since the sources in these regions of the world are much smaller. As can be seen from Table 4-4 and Figure 4-11, nearly 70% of the NO_x sources are anthropogenic, and almost all of them are located in the Northern Hemisphere.

Table 4-4. Tropospheric Sources of NO_x and Their Strengths

	Lower bound	Mean	Upper bound
I. Surface sources			
A. Fossil fuel burning			
Coal	1.9	3.9	5.8
Lignite	.8	1.6	2.3
Light fuel oil	.5	.7	.9
Heavy fuel oil	.7	1.1	1.5
Natural gas	.6	1.9	3.1
Automobiles	3.7	4.3	4.9
Subtotal for A:	8.2	13.5	18.5
B. Soil release	1	5.5	10
C. Biomass burning			
Savanna	1.8	3.1	4.3
Deforestation	.8	2.1	3.4
Fuel wood	1	2	3
Agricultural refuse	2	4	6
Subtotal for C:	5.6	11.2	16.4
Subtotal for surface sources:	14.8	30.2	45.2
II. Atmospheric sources			
NH_3 oxidation	1.2	3.1	4.9
Lightning	2	5	8
High flying aircraft	.2	.3	.4
NO_y transported from stratosphere	.3	.6	.9
Subtotal for atmospheric sources:	3.7	9.0	14.2
III. Total production	19	39	59

Note: The units are 1×10^{12} g N yr^{-1}. The estimates refer to the year 1975 (*from Ehhalt and Drummond, 1982; copyright © 1982 by D. Reidel Publishing Company, Dordrecht, Holland*).

The importance of the abundance of NO_x in the troposphere on the photochemical production of tropospheric ozone is highlighted by Figure 4-12. Another interesting feature of the calculations summarized by this figure is that the amount of ozone photochemically destroyed in the Northern Hemisphere remains relatively constant near 21×10^{10} mol cm^{-2} s^{-1} regardless of the assumed NO_x concentration (Fishman, Solomon, and Crutzen, 1979b). Assuming that the photochemical destruction rate of tropospheric ozone is well defined, one can speculate that the average Northern Hemispheric tropospheric NO_x concentration should be about 0.040 ppbv to balance all the source and sink terms in the derivation of the hemispheric budget. Again, however, it must be emphasized that the tropospheric distribution of NO_x is most likely very heterogeneous and that there should exist regions in which photochemical processes act as a sizeable source of tropospheric O_3. The study of Fishman and Seiler (1983) suggests that one such region may exist in the free troposphere between 15°N and 45°N based on the concurrent measurements of O_3 and CO using fast-response instruments. In their study, Fishman and Seiler noted that most of the short-term fluctuations of these two trace gases were positively correlated in this region of the troposphere and that such a finding would be consistent with the existence of a photochemical source of tropospheric ozone. On the other hand, it is also conceivable

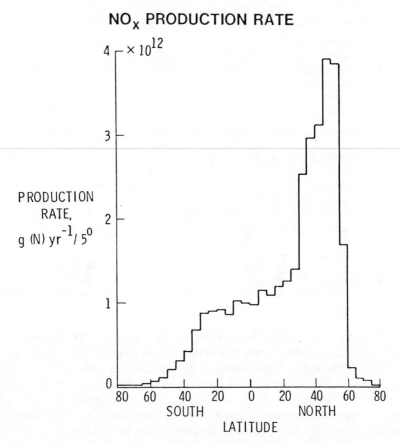

Figure 4-11. Latitudinal comparison of the estimated total production of nitrogen oxides for 1975. The production estimates include both the free-tropospheric and the ground-level sources. *(After Ehhalt and Drummond, 1982, The tropospheric cycle of NO_x, in Chemistry of the Unpolluted and Polluted Troposphere, H. W. Georgii and W. Jaeschke, eds., Reidel, Dordrecht, Holland.)*

that the very low concentrations of tropospheric ozone found in the marine tropical boundary layer (see Routhier et al., 1980; Liu et al., 1983) may be a result of photochemical destruction, since the deposition rate of ozone over ocean surfaces should be very small. The diagnostic models predict that the photochemical loss rate of ozone in the marine tropical boundary layer should be quite large, especially if the very low NO concentrations measured by McFarland et al. (1979) are universally representative of the marine tropical boundary layer. Thus, although there is no conclusive proof that photochemical processes are instrumental for defining the distribution of tropospheric ozone, there are some aspects of the observations that support the important role of photochemistry in the tropospheric ozone budget. As of now, however, the importance of photochemical processes, as suggested by the photochemical modeling experiments, is far from verified.

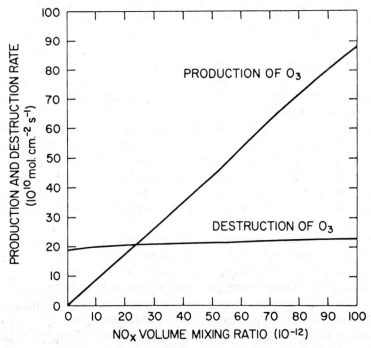

Figure 4-12. Model-derived annual mean photochemical production and destruction rates of tropospheric ozone in the Northern Hemisphere as a function of the prescribed NO_x concentrations. *(From Fishman, J., S. Solomon, and P. J. Crutzen, 1979, Observational and theoretical evidence in support of a significant in-situ photochemical source of tropospheric ozone, Tellus 31.)*

SUMMARY

Although the present-day understanding of the tropospheric ozone cycle incorporates a significant contribution from photochemical processes, it remains very unclear how much influence man's activities have had on the natural component of the tropospheric ozone cycle. One of the main obstacles in the determination of the anthropogenic component of the tropospheric ozone budget is that there are no direct emissions of O_3. Thus, the quantification of the perturbation to the natural tropospheric odd-oxygen cycle must include a good knowledge of the anthropogenic components of the budgets of several precursor trace gases that lead to ozone formation. The most important of these trace gases are NO_x and CO, although the oxidation of methane and other hydrocarbons may contribute significantly to the photochemical production of tropospheric O_3. Ehhalt and Drummond (1982) estimate that two-thirds of the amount of NO_x released to the troposphere is produced from anthropogenic processes, whereas Logan et al. (1981) estimate that nearly half the CO in the atmosphere is present as a result of human activity. For both these species, and for both CH_4 and nonmethane hydrocarbons, the source strengths are considerably larger in the Northern Hemisphere than in the Southern Hemisphere. Therefore, we should find considerably more O_3 in the Northern Hemisphere troposphere than in the Southern Hemisphere if there has been a significant perturbation to the natural tropospheric O_3 budget.

Complicating this argument is the suggestion by several circulation modeling studies that the primary natural source of tropospheric ozone (i.e., transport from the stratosphere) should likewise be considerably larger in the Northern Hemisphere than in the Southern Hemisphere. Thus, it is indeed a complex task to determine the anthropogenic effect on the tropospheric O_3 budget. However, the consequences of such a perturbation are by no means insignificant.

Tropospheric ozone is a strong absorber of infrared radiation at 9.6 μm, and the terrestrial radiative energy balance is quite sensitive to the column-integrated amount of tropospheric ozone (Ramanathan and Dickinson, 1979). The difference between the tropospheric column amounts observed at 40°N and 40°S results in a computed temperature difference of 0.3 K due to the greenhouse effect of only the tropospheric ozone. If the amounts of O_3 in the troposphere were to double in the Northern Hemisphere in the forthcoming years, the effect would be an increase of 0.9 K at the earth's surface (Fishman et al., 1979). Through an analysis of the available ozonesonde data, Logan (1982) concludes that tropospheric O_3 may have increased 10-20% between 1960 and 1980 in the Northern Hemisphere. Based on this trend, the increase in temperature due to the increased amount of tropospheric ozone would be two to three times greater than the calculated greenhouse effect due to the observed increase in CO_2 over the same period of time. In the Southern Hemisphere, no such trend in tropospheric ozone has been detected (see Pittock, 1977).

The atmospheric residence time of tropospheric ozone (on the order of one month) is considerably shorter than the characteristic time for interhemispheric mixing (on the order of one year), so any secular increase in tropospheric ozone in the Northern Hemisphere should result in a climatic perturbation only in that hemisphere. All other radiatively active gases (Hansen et al., 1981) that contribute to the greenhouse effect and that may be increasing due to man's activities (i.e., CO_2, CH_4, N_2O, $CFCl_3$, and CF_2Cl_2) would cause a warming that would be global in nature, since their atmospheric residence times are on the order of a decade or longer. The effect of tropospheric ozone, on the other hand, would be confined to the Northern Hemisphere, which further complicates the consequences of a possible climatic perturbation.

Thus, the attainment of an understanding of the tropospheric ozone cycle will lead to a better knowledge of the biogeochemical atmospheric cycles of both carbon and nitrogen. In addition, the climatic consequences resulting from a perturbation to the natural tropospheric ozone cycle are perhaps more complex and more far-reaching than human-induced perturbations of any other trace gas in the atmosphere.

REFERENCES

Aldaz, L., 1969, Flux measurements of atmospheric ozone over land and water, *J. Geophys. Res.* **74**:6943-6946.

Bollinger, M. J., D. D. Parrish, C. Hahn, D. L. Albritton, and F. C. Fehsenfeld, 1982, NO_x measurements in clean continental air, paper presented at the Second Symposium on the Composition of the Nonurban Troposphere (preprint available from American Meteorological Society, Boston, Mass.)

Chameides, W. L., 1978, The photochemical role of tropospheric nitrogen oxides, *Geophys. Res. Lett.* **5**:17-20.

Chameides, W. L., and D. H. Stedman, 1977, Tropospheric ozone: Coupling transport and photochemistry, *J. Geophys. Res.* **82**:1787-1794.

Chameides, W. L., and A. Tan, 1981, The two-dimensional diagnostic model for OH: An uncertainty analysis, *J. Geophys. Res.* **86**:5209-5224.

Chameides, W. L., and J. C. G. Walker, 1973, A photochemical theory of tropospheric ozone, *J. Geophys. Res.* **78**:8751-8760.

Chameides, W. L., and J. C. G. Walker, 1976, A time-dependent photochemical model for ozone near the ground, *J. Geophys. Res.* **81**:413-420.

Chatfield, R., and H. Harrison, 1976, Ozone in the remote troposphere: Mixing versus photochemistry, *J. Geophys. Res.* **81**:421-423.

Chatfield, R., and H. Harrison, 1977*a*, Tropospheric ozone 1: Evidence for higher background values, *J. Geophys. Res.* **82**:5965-5968.

Chatfield, R., and H. Harrison, 1977*b*, Tropospheric ozone 2: Variations along a meridional band, *J. Geophys. Res.* **82**:5969-5976.

Crutzen, P. J., 1973, A discussion of the chemistry of some minor constituents in the stratosphere and troposphere, *Pure Appl. Geophys.* **106-108**:1385-1399.

Crutzen, P. J., 1974, Photochemical reactions initiated by and influencing ozone in unpolluted tropospheric air, *Tellus* **26**:47-57.

Danielsen, E. F., 1964, *Project Springfield Report,* Report no. DASA 1517, Defense Atomic Support Agency, Washington, D.C.

Danielsen, E. F., 1968, Stratospheric-tropospheric exchange based on radioactivity, ozone, and potential vorticity, *J. Atmos. Sci.* **25**:502-518.

Danielsen, E. F., and V. A. Mohnen, 1977, Project Dustorm report: Ozone measurements and meteorological analysis of tropopause folding, *J. Geophys. Res.* **82**:5867-5878.

Dickerson, R. R., D. H. Stedman, W. L. Chameides, P. J. Crutzen, and J. Fishman, 1979/1980, Actinometric measurements and theoretical calculations of j(O$_3$), the rate of photolysis of ozone to O(^{1}D), *Geophys. Res. Lett.* **6**:833-835 (correction in *Geophys. Res. Lett.* **7**:112).

Ehhalt, D. H., and J. W. Drummond, 1982, The tropospheric cycle of NO$_x$, in *Chemistry of the Unpolluted and Polluted Troposphere,* H. W. Georgii and W. Jaeschke, eds., Reidel, Dordrecht, Holland, 219-252.

Fabian, P., 1973, A theoretical investigation of tropospheric ozone and stratospheric-tropospheric exchange processes, *Pure Appl. Geophys.* **106-108**:1044-1057.

Fabian, P., 1974, Comments on "A photochemical theory of tropospheric ozone" by W. L. Chameides and J. C. G. Walker, *J. Geophys. Res.* **79**:4124-4125.

Fabian, P., and C. E. Junge, 1970, Global rate of ozone destruction at the earth's surface, *Arch. Met. Geoph. Biokl.,* ser. A, **19**:161-172.

Fabian, P., and P. G. Pruchniewicz, 1977, Meridional distribution of ozone in the troposphere and its seasonal variations, *J. Geophys. Res.* **82**:2063-2073.

Fishman, J., and P. J. Crutzen, 1977, A numerical study of tropospheric photo-chemistry using a one-dimensional model, *J. Geophys. Res.* **82**:5897-5906.

Fishman, J., and P. J. Crutzen, 1978, The origin of ozone in the troposphere, *Nature* **274**:855-858.

Fishman, J., and W. Seiler, 1983, The correlative nature of ozone and carbon monoxide in the troposphere: Implications for the tropospheric ozone budget, *J. Geophys. Res.* **88**:3662-3670.

Fishman, J., W. Seiler, and P. Haagenson, 1980, Simultaneous presence of O$_3$ and CO bands in the troposphere, *Tellus* **32**:456-463.

Fishman, J., S. Solomon, and P. J. Crutzen, 1979, Observational and theoretical evidence in support of a significant in-situ photochemical source of tropospheric ozone, *Tellus* **31**:432-446.

Fishman, J., V. Ramanathan, P. J. Crutzen, and S. C. Liu, 1979, Tropospheric ozone and climate, *Nature* **282**:818-820.

Galbally, I. E., and C. R. Roy, 1980, Destruction of ozone at the earth's surface, *R. Meteorol. Soc. Q. J.* **106**:599-620.

Gidel, L. T., and M. A. Shapiro, 1980, General circulation estimates of the net vertical flux of ozone in the lower stratosphere and the implications for the tropospheric ozone budget, *J. Geophys. Res.* **85**:4049-4058.

Hansen, J., D. Johnson, A. Lacis, S. Lebedeff, P. Lee, D. Rind, and G. Russell, 1981, Climatic impact of increasing carbon dioxide, *Science* **213**:957-966.

Hering, W. S., and T. R. Borden, Jr., 1964-1967, *Ozone Observations over North America,* vols. 1-4, AFCRL-64-30, Air Force Cambridge Research Laboratories, Bedford, Mass.

Howard, C. J., and K. M. Evenson, 1977, Kinetics of the reaction of HO_2 with NO, *Geophys. Res. Lett.* **4**:437-440.

Lenschow, D. H., and E. W. Barry, 1980, Photochemical modeling: Aircraft data, *The CHON Photochemistry of the Troposphere,* Notes from a Colloquium, A. C. Delany, ed., (NCAR/CQ-7+1980-ASP, available from the National Center for Atmospheric Research, Boulder, Colo., 114-130.

Lenschow, D. H., R. Pearson, Jr., and B. B. Stankov, 1981, Estimating the ozone budget in the boundary layer by use of aircraft measurements of ozone eddy flux and mean concentration, *J. Geophys. Res.* **86**:7291-7298.

Lenschow, D. H., R. Pearson, Jr., and B. B. Stankov, 1982, Measurements of ozone vertical flux to ocean and forest, *J. Geophys. Res.* **87**:8833-8838.

Liu, S. C., 1977, Possible effects on tropospheric O_3 and OH due to NO emissions, *Geophys. Res. Lett.* **4**:325-328.

Liu, S. C., D. Kley, M. McFarland, J. D. Mahlman, and H. Levy II, 1980, On the origin of tropospheric ozone, *J. Geophys. Res.* **85**:7546-7552.

Liu, S. C., M. McFarland, D. Kley, O. Zafiriou, and B. Huebert, 1983, The photochemistry of near surface air over the equatorial Pacific *J. Geophys. Res.* **88**:1360-1368.

Logan, J. A., 1982, Trends in tropospheric ozone, paper presented at the Second Symposium on the Composition of the Nonurban Troposphere (preprint available from American Meteorological Society, Boston, Mass.)

Logan, J. A., M. J. Prather, S. C. Wofsy, and M. B. McElroy, 1981, Tropospheric chemistry: A global perspective, *J. Geophys. Res.* **86**:7210-7254.

McFarland, M., D. Kley, J. W. Drummond, A. L. Schmeltekopf, and R. H. Winkler, 1979, Nitric oxide measurements in the equatorial Pacific region, *Geophys. Res. Lett.* **6**:605-608.

Mahlman, J. D., H. Levy II, and W. J. Moxim, 1980, Three-dimensional tracer structure and behavior as simulated in two ozone precursor experiments, *J. Atmos. Sci.* **37**:655-685.

Mohnen, V. A., and E. R. Reiter, 1977, International Conference on Oxidants, 1976—*Analysis of Evidence and Viewpoints, Part III. The Issue of Stratospheric Ozone Intrusion,* EPA-600/3-77-115, U.S. Environmental Protection Agency, Washington, D.C.

Noxon, J. F., 1978, Tropospheric NO_2, *J. Geophys. Res.* **83**:3051-3057.

Oltmans, S. J., 1981, Surface ozone measurements in clean air, *J. Geophys. Res.* **86**:1174-1180.

Pearson, R., Jr., and D. H. Stedman, 1980, Instrumentation for fast response ozone measurements from aircraft, *Atmospheric Technology* **12**:50-54 (available from Natl. Cent. Atmos. Res., Boulder, CO).

Pearson, R., Jr., B. F. Weber, D. H. Lenschow, and B. B. Stankov, 1982, Ozone flux and mean concentration budget over short-grass prairie, paper presented at the Second Symposium on the Composition of the Nonurban Troposphere (preprint available from American Meteorological Society, Boston, Mass.

Pittock, A. B., 1974, Ozone climatology, trends, and the monitoring problem, in *Proceedings of the International Conference on Structure, Composition, and General Circulation of the Upper and Lower Atmospheres and Possible Anthropogenic Perturbations,* vol. 1, January 14-25, Melbourne, Australia, 455-466. Cosponsored by the International Association of Meteorology and Atmospheric Physics, The Australian Academy of Sciences, and the U.S. Department of Transportation.

Pittock, A. B., 1977, Climatology of the vertical distribution of ozone over Aspendale (38°S, 145°E), *R. Meteorol. Soc. Q. J.* **103**:575-584.

Ramanathan, V., and R. E. Dickinson, 1979, The role of stratospheric ozone in the zonal and seasonal radiative energy balance of the earth-troposphere system, *J. Atmos. Sci.* **36**:1084-1104.

Robinson, E., and R. C. Robbins, 1969, *Sources, Abundance, and Fate of Gaseous Atmospheric Pollutants,* Standford Research Institute, Menlo Park, Calif.

Routhier, F., and D. D. Davis, 1980, Free tropospheric/boundary-layer airborne measurements of H_2O over the latitude range of 58°S to 70°N: Comparison with simultaneous ozone and carbon monoxide measurements, *J. Geophys. Res.* **85**:7293-7306.

Routhier, F., R. Dennett, D. D. Davis, E. Danielsen, A. Wartburg, P. Haagenson, and A. C. Delany, 1980, Free tropospheric and boundary layer airborne measurements of ozone over the latitude range of 58°S and 70°N, *J. Geophys. Res.* **85**:7307-7321.

Seiler, W., and J. Fishman, 1981, The distribution of carbon monoxide and ozone in the free troposphere, *J. Geophys. Res.* **86**:7255-7266.

Shapiro, M. A., 1980, Turbulent mixing within tropopause folds as a mechanism for the exchange of chemical constituents between the stratosphere and troposphere, *J. Atmos. Sci.* **37**:994-1004.

Viezee, W., W. B. Johnson, and H. B. Singh, 1979, *Airborne Measurements of Stratospheric Ozone Intrusions into the Free Troposphere over the United States,* vol. 1, *Data Analysis and Interpretation,* Final Report, SRI Project 6690, Stanford Research Institute, Menlo Park, Calif.

Wesely, M. L., D. R. Cook, and R. M. Williams, 1981, Field measurement of small ozone fluxes to snow, wet bare soil, and lake water, *Boundary Layer Meteor.* **20**:459-472.

Chapter 5

Stratospheric Ozone Perturbations

R. P. Turco

R & D Associates

Today, there is a major scientific focus on the origins and properties of stratospheric ozone. Interest has been spurred by the discovery that man's activities may significantly alter ozone concentrations from natural levels. The environmental, biological, and climatological consequences of ozone perturbations are discussed in the Introduction and in Chapter 6. This chapter presents a review of the diverse chemical and physical agents that may produce changes in ozone, the analytical methods that are used to study these changes, and current estimates of ozone perturbations that might be caused by natural and artificial means.

Intensive theoretical and experimental investigations of stratospheric ozone have been underway for more than a decade. The aim of the work is to elucidate the formation mechanisms and meteorology of the background ozone layer, the factors that can lead to ozone change, and the biological and climatological implications of ozone alterations. The wide spectrum of natural spatial and temporal fluctuations in ozone concentrations must be taken into account in gauging human impact. In many cases, potential anthropogenic effects must be projected far into the future in order to develop meaningful control and mitigation strategies. Thus, theoretical models—used as numerical forecasting tools—have assumed great importance in assessing threats to the ozone layer. Careful laboratory and field measurements are constantly needed to guide and to validate model development.

It is fitting that several in-depth and critical assessments of the ozone problem have been carried out by atmospheric scientists under the sponsorship of government agencies in several countries (CIAP, 1975; COMESA, 1975; National Academy of Sciences, 1975a, 1975b, 1976, 1979; Hudson, 1977; Hudson and Reed, 1979; U.K. DOE, 1979; WMO/NASA, 1982). These authoritative works provide a historical perspective on stratospheric ozone research.

Tables 5-1 through 5-4 (see pages 198-205) summarize the chemical and physical agents— both natural and anthropogenic—that may contribute to ozone change. All these agents have been studied with theoretical models, as exemplified by the references listed in the tables. Foremost among the ozone-active agents currently under investigation are the nitrogen oxides, NO_x ($NO + NO_2$), and the fluorocarbons, $CFCl_3$ and CF_2Cl_2.

The primary anthropogenic sources of stratospheric NO_x are aircraft engines and the decomposition of N_2O originating in human activities. Johnston (1971) pointed out the possible threat to the ozone layer of exhaust from high-flying supersonic aircraft. He connected the NO_x emissions of airplane engines with potential depletions of stratospheric ozone using laboratory observations of fast reactions between NO_x and ozone (see also Crutzen, 1970; 1971).

In their classic paper, Molina and Rowland (1974a) revealed the possible danger to ozone posed by manufactured fluorocarbons, which are used primarily as refrigerants and aerosol propellants. They pointed out that fluorocarbons decompose in the stratosphere, forming chlorine compounds that catalytically attack ozone.

In these early years of scientific analysis of ozone (prior to 1976), important discoveries and abrupt changes in thinking occurred with regularity. Controversy flourished in this environment, and atmospheric science had more than a few inglorious moments, as chronicled in the "Ozone War" (Dotto and Schiff, 1978). Nevertheless, our understanding of stratospheric chemistry and meteorology has progressed steadily to the point where ozone perturbation calculations can now be made with some confidence.

Of course, the study of ozone perturbations assumes the existence of an unperturbed reference state. Generally, both an observational and a theoretical reference state must be determined. The earlier chapters of this book discuss the observational parameters that establish the ozone layer as a constant and ubiquitous feature of the stratosphere, but they also show that ozone is in a continuous state of evolution in both time and space. Hence, the observational reference state for a specific situation depends on the degree of spatial and temporal averaging that must be carried out. For example, a single globally averaged ozone column abundance (in molecules/cm^2 or cm-atm) over a fixed time interval can be derived using available measurements.

The theoretical reference state is determined using an ozone model in which a large number of photochemical and dynamical parameters and boundary conditions are specified. Changes in ozone are usually calculated by varying a small subset of the physical or chemical parameters. If it has been verified that the properties of the observational and theoretical ozone reference states agree within reasonable bounds, then the perturbed states may be meaningfully compared or, when observations are not available for the perturbed state, the model forecasts of ozone change may be employed with greater confidence.

Before proceeding to a quantitative discussion of ozone perturbations, it is worthwhile to review briefly the analytical models used to estimate ozone change.

OZONE MODELS

Models of atmospheric dynamics, radiation, and photochemistry are predicated on the coupled time- and space-dependent integro-differential equations of atmospheric physics and chemistry. These equations, in their complete form, are extraordinarily complex. Detailed discussions of atmospheric physics and related mathematical analyses can be found in appropriate aeronomy and meteorology textbooks (e.g., Whitten and Poppoff, 1971; Holton, 1979). The full set of equations that apply to stratospheric ozone has never been solved. Thus, all ozone models are based on a number of physical approximations.

Prior to the 1960s, most theoretical studies of the atmosphere were carried out using simple and ingenious mathematical formulations. With the availability of large, fast digital computers in the late 1960s, however, discrete finite-difference models of great sophistication quickly replaced the analytical models. When questions about ozone arose in the early 1970s, therefore, it was logical that computer models were brought to bear on the problem.

One of the most obvious characteristics of an atmospheric model is its spatial dimensionality. A box model, for example, treats an isolated air parcel with regard to its radiation balance and/or photochemical state. Box models are most useful for exploratory investigations of chemical processes. One-dimensional (1-D) models are typically used to investigate vertical columns of stratified air; they crudely take into account the transfer of materials between adjacent levels as a result of turbulent or eddy diffusion and convection. One-dimensional models have been widely applied to analyze the average properties of the global atmosphere and,

Table 5-1. Natural Physical Events Affecting Ozone

Physical Agent	Description	Nature of the Effect[a]	Scenarios for Ozone Change	Researchers
Solar variations	Natural long-term cyclic variations in the solar ultraviolet output (e.g., the 11-year cycle).	Increased uv-fluxes lead to greater ozone (odd-oxygen) production by O_2 photodissociation. Ozone concentrations covary with the solar uv flux.	Satellite ozone observations have revealed 11-year and 37-day solar cycle variations in ozone above ~30 km.	Ruderman and Chamberlain (1975) Callis and Nealy (1978) Penner and Change (1978, 1980) Keating et al. (1981)
	Diurnal variations.	Daily modulation of molecular photolysis rates.	Small effects on stratospheric ozone concentrations have been predicted and observed.	Whitten and Turco (1974a) Vaughan (1982)
	Solar eclipses.	Short-term perturbation of ozone photochemistry.	No discernible effects on stratospheric ozone.	Herman (1979) Wuebbles and Chang (1979)
Solar proton events (SPEs)	Protons emitted during solar storms penetrate the Earth's upper atmosphere.	Energetic protons deposited in air generate NO_x and HO_x which catalytically react with ozone.	The large SPE of August 1972 caused a maximum ozone depletion of ~20% at 40 km, but a total ozone column change <2%.	Heath et al. (1977) Fabian et al. (1979) Reagan et al. (1981) McPeters et al. (1981)
Aurorae	Electrons in the earth's radiation belts precipitate into the upper atmosphere.	High energy electrons generate NO_x in the mesosphere, which may be transported to the stratosphere.	The contribution of auroral NO_x to natural ozone loss is probably <5%.	Bauer (1979) Solomon et al. (1982) Thorne (1977)
Meteors	Continual accretion of interplanetary dust, and occasional influx of asteroidal or cometary debris.	Aerodynamic heating of air by high-velocity meteors generates NO_x, which can react with ozone.	Small meteoroids are stopped in the mesosphere and have a negligible effect on stratospheric ozone. Large meteors (e.g., Tunguska) can	Park and Menees (1978) Park (1978) Turco et al. (1981a, 1982a)

		reach the stratosphere and cause large ozone reductions, up to 30%; such events are rare (one each 1-20 millenia). The ozone layer recovers in ~3-5 years.	
Supernovae	Terminal stellar explosions.	Emitted energetic particles and radiation penetrate the stratosphere, and generate NO_x which reduces ozone.	Whitten et al. (1976) Reid et al. (1978) Aikin et al. (1980)
		For supernovae within ~100 light-years, the cosmic rays could deplete ozone by 30 to 80%. Such events are exceedingly rare—one each 100 million years. Ozone depletions persist for centuries.	
Volcanoes	Continuous natural venting of heat from the Earth's interior. Large Plinian eruptions inject material well above 15 km altitude.	Volcanic injections of H_2O and HCl can reduce ozone chemically (see the entries for H_2O and HCl). Large injections of sulfur gases (SO_2 and H_2S) can suppress the HO_x cycle. Dense volcanic particle layers can alter photodissociation rates.	Fiocco et al. (1978) Stolarski and Butler (1979) Cadle (1980) Turco et al. (1982b, 1983)
		For large eruptions (e.g., Agung) ozone in the hemisphere of eruption might be reduced by 3% or less. For moderate eruptions (e.g., Mount St. Helens) ozone reductions <1% are expected. The ozone layer recovers in 2-3 years.	

[a] Only the primary stratospheric effects are summarized. For a description of the effects caused by secondary chemical agents such as NO_x and HO_x, see the other tables in this series. Chapter 2 presents a detailed review of ozone photochemical reaction mechanisms. Possible effects on tropospheric ozone are discussed in Chapter 4, and related climate variations, in Chapter 6.

Table 5-2. Nitrogen Compounds Affecting Ozone

Chemical Agent	NO_x Sources (*natural)	Nature of the Effect[a]	Scenarios for Ozone Change[b]	Researchers
Nitrous oxide (N_2O)	*Decomposition of biogenic and auroral N_2O by $O(^1D)$ yielding NO_x (also see Table 5-1: Solar Proton Events, Meteors, and Aurorae).	NO and NO_2 react catalytically with O and O_3: $NO + O_3 \rightarrow NO_2 + O_2$ $\underline{NO_2 + O \rightarrow NO + O_2}$ (net) $O + O_3 \rightarrow O_2 + O_2$ Large NO_x increases can lead to radiation/temperature perturbations.	Background NO_x is believed to cause about 30–40% of the natural ozone loss in the stratosphere.	Bates and Hays (1967) McElroy and McConnell (1971) Crutzen (1974a) Nicolet (1975) Johnston (1977) Zipf and Prasad (1980)
	Anthropogenic N_2O decomposition by $O(^1D)$.	See above.	Doubling of background N_2O could lead to a 10–16% reduction in total ozone.	McElroy et al. (1976) Crutzen (1976) Liu et al. (1976a) Sze and Rice (1976) Johnston (1977) Whitten et al. (1981, 1983)
Nitrogen oxides (NO_x)	Supersonic aircraft (SST) exhaust.	NO/NO_2 engine emissions react with O_3, as described above.	A commercial fleet of SST's operating at ~20 km emitting ~1×10^9 kg NO_2/yr might cause a 4–8% ozone column reduction.	Johnston (1971) Crutzen (1972) Schiff and McConnell (1973) McElroy et al. (1974) Whitten and Turco (1974b) National Academy of Sciences (1975a) CIAP (1975)

Source			References
Subsonic aircraft exhaust.	No$_x$ emissions below 20 km can interfere with the HO$_x$–O$_3$ reaction sequence (see H$_2$O), leading to small net ozone gains.	Commercial air traffic operating up to ~13 km emitting ~2×10^9 kg NO$_2$/yr may induce a −1 to +2% change in the ozone column.	CIAP (1975) Broderick (1977) Poppoff et al. (1978)
Nuclear explosions.	Air heated above ~3000 K in a fireball produces ~10^{32} NO molecules per megaton of explosive energy.	Nuclear tests of the 1950s and 1960s may have reduced ozone in the Northern Hemisphere by 1-3% for about 1-2 years. A 10,000 megaton nuclear war depositing ~10^{36} NO molecules in the stratosphere could lead to a maximum N.H. O$_3$ reduction of 50-70%, with recovery in ~3-5 years.	Johnston et al. (1973) Hampson (1974) National Academy of Sciences (1975b) Whitten et al. (1975a) MacCracken and Chang (1975) Chang et al. (1979) Crutzen and Birks (1982)

[a] Only the primary stratospheric effects are given. Possible effects on tropospheric ozone are discussed in Chapter 4, and related climate variations in Chapter 6. Chapter 2 presents a detailed review of ozone photochemical reaction mechanisms. Also note that the various chemical cycles (e.g., O$_x$, HO$_x$, NO$_x$, Cl$_x$, Br$_x$, etc.) are closely coupled, and perturbations to one cycle may strongly affect the other cycles. Complex scenarios involving the coupled effects of two or more agents acting simultaneously are considered in the text.

[b] Steady-state ozone perturbations are given unless otherwise stated.

Table 5-3. Halogen Compounds Affecting Ozone

Chemical Agent	Sources (*natural)	Nature of the Effect[a]	Scenarios for Ozone Change[b]	Researchers
Chlorofluoromethanes (CFMs)	Manufactured $CFCl_3$ and CF_2Cl_2, which diffuse into the stratosphere.	Photodecomposition of CFM's near 30 km generates Cl_x (Cl and ClO) which reacts catalytically with O and O_3: $$Cl + O_3 \rightarrow ClO + O_2$$ $$ClO + O \rightarrow Cl + O_2$$ (net) $O + O_3 \rightarrow O_2 + O_2$	CFM release to the atmosphere at current production rates could, in 50-100 years, reduce global ozone by 3-10%. Recovery of the ozone layer, following an end to CFM emission, might take ~50-150 years.	Molina and Rowland (1974a) Cicerone et al. (1974, 1975) Crutzen (1974b) Wofsy, McElroy, and Sze (1975) Rowland and Molina (1975) Turco and Whitten (1975) NAS (1976, 1979) Hudson (1977) Hudson and Reed (1979) Borucki et al. (1980) WMO/NASA (1982)
	Chlorocarbons $CHCl_3$, CCl_4, CH_3CCl_3, etc. of industrial origin.	Same as above, except that compounds with C-H bonds are also subject to decomposition by reaction with OH.	Current industrial emissions may eventually cause a 1-3% decrease in total ozone.	Molina and Rowland (1974b) Yung et al. (1975) Crutzen et al. (1978) Logan et al. (1978) McConnell and Schiff (1978) Cicerone (1981)
	Fluorocarbon substitutes such as $CHClF_2$.	Because of decomposition in the troposphere, these compounds produce only ~one-fifth the effect of $CFCl_3$ and CF_2Cl_2.	The future impact on ozone depends upon the extent of utilization of substitute fluorocarbons.	Hudson and Reed (1979) WMO/NASA (1982)

Agent	Source	Mechanism	Effect	References
Hydrogen chloride (HCl)	*Decomposition of natural chlorine compounds of tropospheric origin (e.g., CH_3Cl). Also see Volcanoes in Table 5-1.	HCl reacts with OH releasing Cl, which reacts with ozone as noted above for CFMs.	Background (nonanthropogenic) HCl is thought to cause <5% of the natural ozone loss in the stratosphere.	Stolarski and Cicerone (1974) Wofsy and McElroy (1974) Hudson and Reed (1979)
	Solid-fueled rocket motor exhaust.	See above.	Fifty space shuttle launches per year release enough HCl to reduce total ozone by ~0–0.2%. Recovery of the ozone layer occurs in ~3 years.	Stolarski and Cicerone (1974) Whitten et al. (1975b) Potter (1978)
Bromine	*Decomposition of natural brominated compounds (e.g., CH_3Br).	Bromine gas reacts catalytically with ozone, even in the absence of sunlight: $$2(Br + O_3 \rightarrow BrO + O_2)$$ $$BrO + BrO \rightarrow Br + Br + O_2$$ $$\text{(net) } 2O_3 \rightarrow 3O_2$$ Bromine can also interact synergistically with chlorine in consuming O_3.	Background bromine contributes <5% to the total ozone loss in the stratosphere. Anthropogenic bromine levels of only 100 pptv could reduce the stratospheric ozone column by 5–10%.	Wofsy, McElroy, and Yung (1975) Hudson and Reed (1979) Yung et al. (1980) WMO/NASA (1982)

[a] Only the primary stratospheric effects are given. Possible effects on tropospheric ozone are discussed in Chapter 4, and related climate variations, in Chapter 6. Chapter 2 presents a detailed review of ozone photochemical reaction mechanisms. Also note that the various chemical cycles (e.g., O_x, HO_x, NO_x, Cl_x, Br_x, etc.) are closely coupled, and perturbations to one cycle may strongly affect the other cycles. Complex scenarios involving the coupled effects of two or more agents acting simultaneously are considered in the text.

[b] Steady-state ozone perturbations are given unless otherwise stated.

Table 5-4. Miscellaneous Compounds Affecting Ozone

Chemical Agent	Sources (*natural)	Nature of the Effect[a]	Scenarios for Ozone Change[b]	Researchers
Water vapor (H_2O)	*Tropospheric H_2O, injection through the Hadley cell cold trap (also see Volcanoes in Table 5-1).	H_2O reacts with $O(^1D)$ to form HO_x (OH and HO_2) which catalytically attacks ozone: e.g., $$OH + O_3 \rightarrow HO_2 + O_2$$ $$HO_2 + O_3 \rightarrow OH + 2O_2$$ (net) $2O_3 \rightarrow 3O_2$ Water vapor also has a role in the thermal balance of the lower stratosphere.	Background H_2O is thought to contribute about 30–40% to the natural ozone loss rate.	Brewer (1949) Bates and Nicolet (1950) Hudson (1977) Hudson and Reed (1979)
	High-altitude aircraft exhaust.	Water emitted as a combustion byproduct.	An SST fleet operating at ~20 km emitting ~2×10^{11} kg H_2O yr might cause a 1–2% ozone column decrease. Aircraft operating at higher altitudes produce larger effects.	Harrison (1970) Grobecker et al. (1974) CIAP (1975) Poppoff et al. (1978)
	Liquid-fueled H_2/O_2 rocket motors.	Water emitted as a combustion byproduct.	Future fleets of heavy-lift launch vehicles emitting ~1.6×10^9 kg H_2O/yr might cause a <0.01% total ozone reduction.	DOE (1980) Whitten et al. (1982)

204

Methane (CH_4)	*Oxidation of methane of bio-genic origin generates water vapor in the stratosphere.	CH_4 oxidation may contrib-ute up to 50% of the back-ground H_2O above 20 km. CH_4 decomposition produces some HO_x directly.	Singer (1971) Wofsy, McConnell, and McElroy (1972) Chameides et al. (1977) Sze (1977) Owens et al. (1982)
		Alterations in the CH_4 emis-sion rate or lifetime can alter stratospheric composition, particularly due to the reaction of CH_4 with Cl atoms. Methane initiates smog chemistry.	
Carbon dioxide (CO_2)	*Biomass burning, respira-tion, organic decomposition.	CO_2 infrared absorption and emission is an important factor determining the tem-perature structure and dynamics of the troposphere and stratosphere.	Callis and Nealy (1977) Luther et al. (1977) National Research Council (1979) Groves and Tuck (1980)
		An increase in CO_2 causes the stratosphere to cool. The temperature dependences of ozone reactions are such that cooling leads to an ozone enhancement.	
	Fossil fuel combustion, agricultural fires.	See above.	CEQ (1981)
		Doubling of CO_2 concentra-tions might cause a total ozone increase of 3–9%. The effect could persist for many centuries.	

[a] Only the primary stratospheric effects are given. Possible effects on tropospheric ozone are discussed in Chapter 4, and related climate variations, in Chapter 6. Chapter 2 presents a detailed review of ozone photochemical reaction mechanisms. Also note that the various chemical cycles (e.g., O_x, HO_x, NO_x, Cl_x, Br_x, etc.) are closely coupled, and perturbations to one cycle may strongly affect the other cycles.
[b] Steady-state ozone perturbations are given unless otherwise stated. Complex scenarios involving the coupled effects of two or more agents acting simultaneously are considered in the text.

under certain appropriate conditions, the local properties. The major advantage of 1-D models lies in their ability to handle large photochemical reaction systems with modest computer resources. One-dimensional climate models are also frequently employed to predict atmospheric temperature changes associated with ozone perturbations.

Two-dimensional (2-D) models take into account the average meridional motions of the atmosphere as well as the vertical motions by using zonally averaged wind fields and parameterized diffusion coefficients to compensate for nonadvective, oscillatory, and subscale transport components. Modern 2-D models include relatively detailed photochemical and radiation treatments, but they are still expensive and burdensome to use. For simulation purposes, however, they currently represent a useful compromise between 1-D and 3-D (three-dimensional) models (for example, see Whitten et al., 1977; Harwood and Pyle, 1977; Hidalgo and Crutzen, 1977; Widhopf, Glatt, and Kramer, 1977; Brasseur and Bertin, 1978; Vupputuri, 1979; Derwent and Eggleton, 1981; Miller et al., 1981). The spatial averaging implied in 1-D and 2-D models is discussed in Chapter 3.

Three-dimensional models are based on the primitive transport equations for a spherical, rotating, stratified atmosphere. The fidelity of 3-D models in reproducing the meteorology of the atmosphere is continually improving. However, ozone photochemistry has not yet been satisfactorily incorporated into 3-D models, although several preliminary case studies of ozone formation and dispersion have been carried out (Cunnold et al., 1975; Cunnold, Alyea, and Prinn, 1980).

An excellent review of the structure and characteristics of atmospheric models relevant to stratospheric ozone studies is given in the recent WMO/NASA Report (1982). Figure 5-1 summarizes schematically the physical and chemical elements that may be included in stratospheric models and the coupling between the elements.

The precision of a model in representing the atmosphere depends to a large degree on the depth of the treatment of basic chemical and physical processes. It follows that unknown or missing factors caused by incomplete knowledge of the stratosphere can lead to misrepresentation. Moreover, the full coupling of radiation, dynamics and photochemistry indicated in Figure 5-1 is seldom achieved. Often the coupling is accomplished only by compromising one of the elements; for instance, by using truncated photochemical schemes in 2-D and 3-D models. It is also important to note that dynamical processes on horizontal scales of less than about 100 km are rarely treated properly in atmospheric models. Accordingly, simulations invariably yield only approximate results and must be constantly and diligently validated against observational data. The need for timely environmental assessments has led to the general acceptance and use of these imperfect models.

An important uncertainty in current ozone models relates to the accuracy and completeness of the photochemical schemes employed. In the past, major revisions in key rate coefficients, and discoveries of new chemical processes, have led to major adjustments in predictions of ozone perturbations. There is no way to estimate a priori the probability of large systematic errors of this type. Rough estimates of the residual random errors in current stratospheric ozone calculations lie in the range of a factor of two to three (e.g., Hudson, 1977; Hudson and Reed, 1979).

Models of the atmosphere are characterized by crude spatial resolution. Small-scale fluctuations and variability cannot be accurately simulated. Because stratospheric measurements are typically carried out over limited temporal and spatial domains, direct comparisons with model predictions are often difficult and ambiguous, which in turn complicates the process of model validation. The application of initial and boundary conditions to models of stratospheric composition presents another, less critical, problem in ozone simulation.

Interest in forecasting potential ozone changes has led scientists to project pollution scenarios far into the future. However, since no one can foretell complex economic and social events, considerable uncertainty is associated with predictions of future ozone levels. Moreover, because all the chemical and physical agents that affect ozone vary simultaneously, and because their quantitative effects are strongly coupled, ozone forecasts should be based on multiple-agent perturbation scenarios. Such an approach has only recently been undertaken (e.g., WMO/NASA, 1982).

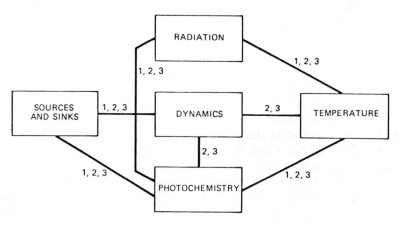

Figure 5-1. The major physical/chemical elements of present-day stratospheric ozone models are illustrated schematically. Important interactions between elements are indicated by solid lines. The arabic numerals indicate the dimensionality of the models in which the elements and couplings between elements have been treated.

A reasonable strategy for studying ozone is to consider a range of perturbation scenarios to determine the possible limits of ozone variation. In this regard, it is prudent to avoid scenarios that project exponentially increasing pollution levels far into the future, as has been done on occasion in the past.

NATURAL OZONE VARIATIONS

There are a large number of geophysical factors that affect natural ozone levels (e.g., Bauer, 1979). Some of these factors are described in Table 5-1. Ozone variations are caused by changes in the rates of photochemical processes and by transport associated with meteorological disturbances, for example, stratospheric warming events (Hilsenrath, 1980; Dütsch and Braun, 1980). The meteorology of stratospheric ozone is discussed in Chapter 3 (also see Ghazi, 1974; Angell and Korshover, 1973, 1976, 1978; London, Frederick, and Anderson, 1977, Hilsenrath, Heath, and Schlesinger, 1979; Frederick et al., 1980; London and Angell, 1982).

Solar-induced Variations

Ultimately, the sun is the source of the ozone layer. Ultraviolet solar radiation photodissociates molecular oxygen leading to the formation of ozone. Likewise, sunlight produces the active chemical species that react with and destroy ozone. In fact, a broad spectrum of the solar electromagnetic and corpuscular emission plays a role in the ozone cycle, from visible and ultraviolet light, which affect atmospheric photochemistry, temperatures, and dynamics, to energetic solar protons, which generate NO_x in the upper atmosphere. It follows that ozone concentrations are sensitive to changes in a variety of solar emissions.

Natural cyclic variations in the solar radiation output, which are particularly marked in the ultraviolet (uv) wavelength region, are well known (Heath, 1973; Cook, Brueckner, and Van, 1980; Brueckner, 1981). Connections between the 11-year sunspot cycle and ozone changes have been found in observational data (Keating, 1978; Blackshear and Tolson, 1978; Dütsch, 1979; London and Reber, 1979; Keating et al., 1981; Reber and Huang, 1982, Chakrabarty and Chakrabarty, 1982). Figure 5-2 shows a theoretical estimate of the maximum variation in ozone, and several other trace species, between solar maximum and solar minimum due to changes in uv fluxes (Callis and Nealy, 1978; Penner and Chang, 1980). The ozone abundance is predicted to vary directly with the solar ultraviolet output, as might be expected. However, the overall variation in the total ozone column is only a few percent, and has not yet been identified unambiguously in the observational record.

Solar proton events (SPEs) periodically deposit large fluxes of high-energy protons in the upper stratosphere. The resultant bombardment of the air generates NO_x and HO_x, which can deplete ozone (Heath, Krueger, and Crutzen, 1977; Fabian, Pyle, and Wells, 1979; Reagan et al., 1981; McPeters, Jackman, and Stassinopoulos, 1981). Figure 5-3 compares satellite observations of high-altitude ozone concentrations against detailed model simulations for the August 1972 SPE (Reagan et al., 1981). The agreement is generally good, although some significant discrepancies exist (McPeters, Jackman, and Stassinopoulous, 1981). Even though peak ozone depletions of $\sim 20\%$ were observed at 42 km, the total ozone column reduction was only $\sim 2\%$ and was not detected.

The diurnal cycle is an important time-varying solar influence. Recent experimental determinations of the diurnal changes in ozone in the upper stratosphere and mesosphere agree quite well with theoretical predictions of the changes (Aimedieu, Rigaud, and Barat, 1981; Wilson and Schwartz, 1981; Vaughan, 1982; Lean, 1982; Fabian, Pyle, and Wells 1982). An example of these comparisons is given in Figure 5-4. Note,

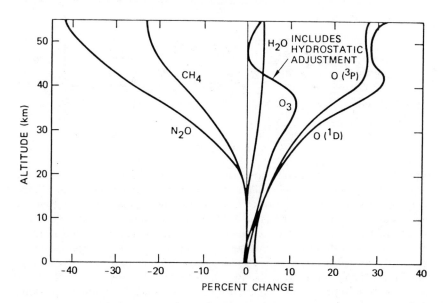

Figure 5-2. Modulation of trace gas concentrations due to the eleven-year cycle in solar ultraviolet emissions. The maximum uv intensity variation is $\pm 15\%$ below 260 nm. The changes in species concentrations are calculated between solar maximum and solar minimum, relative to solar minimum. *(After Penner, J. E., and J. S. Chang, 1980, The relation between atmospheric trace species variabilities and solar uv variability, J. Geophys. Res.* **85.**)

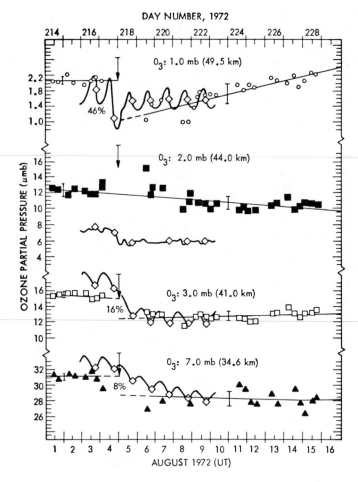

Figure 5-3. Calculated and observed ozone variations at several altitudes for the August 1972 solar proton event (SPE). The calculations, indicated by heavy lines, include diurnal and SPE variations for 77°N latitude. The open diamonds are the predicted ozone concentrations at 0600 Alaska standard time. The Nimbus 4 backscatter ultraviolet data, given by circles, squares and triangles, are averaged over 75°–79°N latitude at the altitudes indicated and correspond to a local time of ~0600 each day (but at different longitudes). Light lines show the average trend in the data. The arrows (roughly) mark the onset of the SPE. The Nimbus 4 data have been reprocessed recently (McPeters, Jackman, and Stassinopoulous, 1981) improving the agreement at 2.0 mb. *(From Reagan et al., 1981, Effects of the August 1972 solar particle events on stratospheric ozone, J. Geophys. Res.* **86**; *copyright © 1981 by the American Geophysical Union.)*

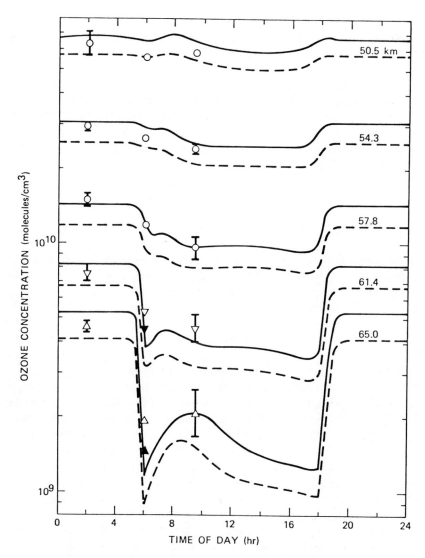

Figure 5-4. Comparison of observed and predicted diurnal ozone variations. Solid lines are theoretical calculations with 2 ppmv of water vapor, and the dashed lines, with 4 ppmv of water vapor. The symbols (open and closed) give the ozone concentrations retrieved from ultraviolet attenuation measurements from rockets. *(After Vaughan, 1982, Diurnal variation of mesospheric ozone, Nature **296.**)*

however, that the diurnal change in the total ozone column is less than 1%. The treatment of diurnal variations in ozone models typically involves an appropriate sunlight averaging scheme (e.g., Turco and Whitten, 1978).

Solar influences on ozone may include less direct mechanisms involving long-term solar modulation of the galactic cosmic ray flux (Ruderman and Chamberlain, 1975; Ruderman, Foley, and Chamberlain, 1976) and solar-storm-triggering of auroral and relativistic electron precipitation events (Thorne, 1977). The impact of these phenomena on ozone is still rather controversial.

Cosmic Perturbations

Cosmic events occasionally cause large atmospheric perturbations. For example, solar eclipses can blot out the sun temporarily on otherwise clear days. Measurements of variations in excited molecular oxygen, $O_2(^1\Delta_g)$, emissions during solar eclipses confirm an expected ozone increase at mesospheric altitudes (Agashe and Rathi, 1982). However, the ozone perturbation caused by an eclipse is generally small and transient (Starr et al., 1980).

On rare occasions, a large meteor or cometary body collides with the earth. If the body is friable, like a comet, a large part of its kinetic energy can be deposited in the atmosphere. The resultant heating of the air along the flight path generates NO_x that subsequently reacts with ozone. Figure 5-5 provides an estimate of the possible ozone perturbations caused by the famous Tunguska meteor fall of 1908 (Turco et al., 1981a, 1982a). Observational evidence pointing to a Tunguska ozone depletion has been uncovered in archived data of the solar spectral intensity recorded at Mount Wilson, California from 1908 to 1911, owing to the distinct O_3 Chappuis band absorption signature in those spectra. On the cosmic scale, Tunguska was a small event. More powerful and less frequent collisions would cause correspondingly larger atmospheric disturbances (Lewis et al., 1982; Turco et al., 1982a).

A spectacular supernova explosion heralds the final epoch in the life of a star. If a supernova were to occur within about 10 parsec (~300 trillion kilometers or 100 light-years) of earth, our atmosphere would be bathed in intense ultraviolet, gamma, and cosmic ray emissions. Several researchers have predicted the changes to be expected in atmospheric composition and ozone concentrations (Whitten et al., 1976; Reid, McAfee, and Crutzen, 1978; Aikin, Chandra, and Stecher, 1980). Aikin, Chandra, and Stecher

(1980) showed that the prompt ultraviolet light might increase total ozone by ~30%, due to enhanced photodissociation of O_2. Whitten et al. (1976) and Reid, McAfee, and Crutzen (1978) considered the delayed cosmic radiation and estimated a range of ozone depletions of 20-90% due to NO_x production. The depletions could persist for 1000-10,000 years and cause major alterations in atmospheric heating rates, meteorology and climate. Fortunately, nearby supernovae are thought to occur only once every 100 million years or so (Reid, McAfee, and Crutzen, 1978).

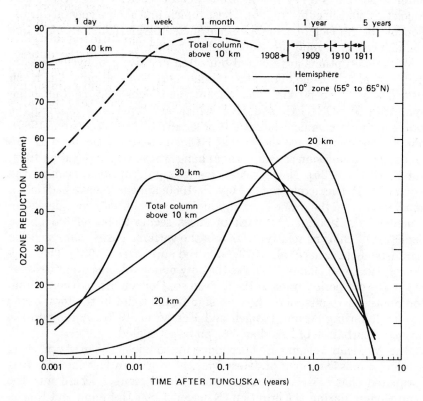

Figure 5-5. Ozone perturbations calculated for the Tunguska meteor event of 1908. Time-varying ozone depletions at several altitudes, and the total ozone column reduction (above 10 km), are shown. The simulation corresponds to equinoctial conditions at 50°N latitude. Two cases of initial horizontal spreading of the meteor-generated NO_x are depicted (over the Northern Hemisphere, and over a 10° latitude zone). *(After Turco et al., 1981a, The Tunguska meteor fall of 1908: Effects on stratospheric ozone, Science **214**.)*

Earth-Induced Variations

The geochemical and biochemical cycles of the earth affect the composition of the atmosphere and the resultant level of ozone in the stratosphere. In this regard, the evolution of the terrestrial atmosphere and the appearance of life over the past 4 billion years has been a subject of intense research (Berkner and Marshall, 1967; Walker, 1977; Pollack and Yung, 1980). Related studies have dealt with the photochemical origins of ozone (Blake and Carver, 1977; Kasting and Donahue, 1980). The hydrological cycle of the atmosphere is also coupled with the ozone cycle inasmuch as water vapor plays a key role in ozone photochemistry (e.g., Bates and Nicolet, 1950, Liu et al., 1976b). These topics, and a more general discussion of the impact of synoptic-scale meteorological factors on ozone abundances, is beyond the scope of this chapter.

Volcanic eruptions are isolated but often massive geophysical events that can affect ozone. Powerful Plinian eruption columns can reach altitudes of 30 to 50 km. The volcanic emissions may contain large quantities of SO_2, H_2O, and HCl, which are capable of altering the ambient ozone chemical balance (Cadle, 1975, 1980; Johnston, 1980). Simulation studies have shown that HCl is potentially the most important volcanic emission causing ozone change (Stolarski and Butler, 1979; Turco et al., 1982b). Nevertheless, many large eruptions, such as the Mount St. Helens explosion of May 18, 1980, appear to have had only a small impact on the ozone layer, producing less than 1% ozone depletion (Turco et al., 1982b). This lack of effect seems to be related to the depletion of chlorine relative to SO_2 in the volcanic gases that reach the stratosphere (Lazrus et al., 1979; Gandrud and Lazrus, 1981). The eruption of Agung on Bali in 1963 was the only major volcanic event to occur in the 70-year period prior to 1982. Even so, Dobson ozone records from the Southern Hemisphere show no significant trend in the total ozone burden following Agung (Angell and Korshover, 1976). A limit to the ozone perturbation of less than 3% can be estimated.

The Mexican volcano El Chichon erupted violently in April, 1982 sending a massive cloud of debris over the Northern Hemisphere. It is estimated that ~3-5×10^6 metric tons of SO_2 were injected into the stratosphere during the eruption (Krueger, 1983; Hofmann and Rosen, 1983). It is also estimated that 0.04 million tons of HCl were injected (Mankin and Coffey, 1984). Computer simulations of the evolving volcanic clouds suggest that ozone could be depleted by 5-10% in the early cloud and by 1-2% over the Northern Hemisphere after a year (Turco et al., 1982c). However, owing to the optical interference caused by volcanic particulate debris in ozone sensors, the predicted changes in ozone would be difficult to detect by remote means.

ANTHROPOGENIC OZONE PERTURBATIONS

During the past decade it has become increasingly apparent that the upper atmosphere can be modified by a number of manmade compounds. Of central concern to aeronomers has been the possible depletion of ozone by these compounds both at their present levels and at levels projected for the future. The perceived danger of the ozone problem has, in fact, motivated much of the current research in atmospheric chemistry and physics. Here, estimates of man's possible impact on ozone are reviewed.

For the most part, ozone predictions have been made with 1-D models, although some limited 2-D calculations are available for comparison. The computational results are generally presented as a steady-state perturbation for a fixed rate of pollutant injection, with all other parameters held constant. Of course, each perturbation exhibits a characteristic time to reach steady state and, if the pollutant source is removed, another characteristic time to decay. The characteristic times depend, in varying degrees, on the pollution scenario (e.g., the rates and heights of injection), the pollutant photochemistry, and the time-scales of atmospheric transport. Many anthropogenic pollutants have natural counterparts in geochemical or biochemical cycles. Moreover, the atmosphere has several means of removing trace materials including photochemical destruction, transport by winds and turbulence, precipitation scavenging (washout and rainout), and absorption at plant, soil, and water surfaces. For each of the perturbations discussed below, ozone would recover to its ambient level after the perturbing factor.

Nitrogen Oxides (NO_x)

Sources of anthropogenic NO_x in the stratosphere are varied. The engines of high altitude aircraft emit NO_x directly as a combustion byproduct. Rocket motors similarly deposit NO_x in launch plumes. Large artificial objects, such as the space shuttle, which periodically re-enter the earth's lower atmosphere from space, produce NO_x by frictional heating of the air in the same way as meteors. Nitrogen oxides are also generated during the photochemical decomposition of anthropogenic N_2O, which is released near the ground and transported into the stratosphere. Finally, large nuclear explosions produce fireballs rich in NO_x which may rise under buoyant forces into the lower stratosphere.

To place NO_x-ozone perturbations in perspective, it should be noted that the global NO_x budget of the stratosphere is $\sim 2 \times 10^8$ NO_x molecules/cm^2-s, or ~ 1 million tons of N per year. The stratosphere holds roughly 3 million tons of fixed- (or odd-) N. Background NO_x contributes ~ 30-40% to the total natural chemical loss of stratospheric ozone (Johnston and Podolske, 1978).

Aircraft and Rockets. Commercial aircraft traffic, both subsonic and supersonic, deposits NO_x in the lower stratosphere. A major interest has developed in the pollution that might be caused by future supersonic transport (SST) fleets, because of their high altitude of operation. Johnston (1971) alerted the scientific community to the possibility of large ozone depletions accompanying the emission of NO_x by SSTs. Figure 5-6 illustrates the history of the subsequent predictions of SST-fleet effects on total ozone (note that the NO_x emission used as a standard measure, 1×10^9 kg NO_2/yr at 20 km averaged over the globe, corresponds to a fleet of several hundred large "advanced" supersonic aircraft; see CIAP, 1975.

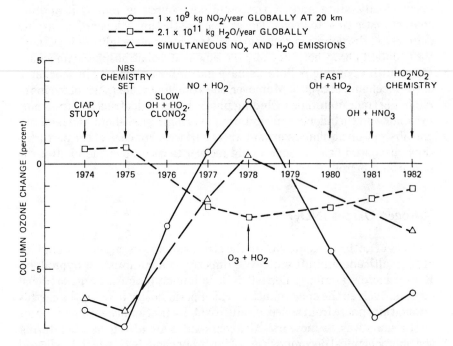

Figure 5-6. Historical trends in model predictions of ozone perturbations caused by SST emissions. Steady-state globally-averaged ozone changes for fixed emission rates of NO_2 and H_2O are given. Not all the significant changes in photochemistry that occurred between 1971 and 1983 are shown. The calculated ozone perturbations are representative of each period. *(From Turco et al., 1978, 1981b, Aircraft NO_x Emissions and Stratospheric Ozone Reductions: Another Look, AIAA Paper 81-0240, Aerospace Sciences Meeting, St. Louis, Missouri, January 12-15.)*

The projected SST effect has changed dramatically over the last decade from initially large decreases in the early 1970s to small *increases* in the late 1970s to large decreases again in the early 1980s. These adjustments in ozone predictions have resulted primarily from modifications in key chemical rate coefficients (see Chapter 2). As knowledge of ozone photochemistry has improved, confidence in ozone predictions has increased. Nevertheless, in the past, large changes in individual, sometimes obscure, chemical reaction rates have led to major changes in ozone forecasts. We must, therefore, be very cautious of model results unconfirmed by field measurements.

The effect on ozone of stratospheric NO_x injections by SSTs decreases with a decreasing height of injection for altitudes below ~30 km (e.g., Poppoff et al., 1978). For injections near the tropopause (as with the present subsonic long-haul commercial jet aircraft fleet), the total ozone column may actually increase by a fraction of a percent, although model uncertainties are too large to preclude small ozone decreases (WMO/NASA, 1982). The altitude region of primary ozone depletion by SST-NO_x injections occurs between about 20 and 30 km. The ozone depletion is not proportional to the NO_x injection; the incremental decrease tends to be larger with each additional unit of NO_x injection.

Figure 5-7 illustrates the possible latitude dependence of the SST ozone perturbation based on two-dimensional model calculations (Borucki et al., 1976; Whitten et al., 1981). Notice that the largest ozone reductions are predicted at high latitudes. The seasonal variation at a fixed latitude is also projected to be smaller than the overall latitudinal variation (Whitten et al., 1981). Further details concerning multidimensional SST simulations are given by Alyea, Cunnold, and Prinn (1975), Widhopf, Glatt, and Kramer (1977) and Hidalgo and Crutzen (1977).

Rocket launches and re-entries made for scientific, commercial, and military purposes are now, and should remain for many decades, a minor source of stratospheric NO_x. For example, proposed space shuttle activity (~50 launches/year) causes only transient, localized perturbations in mesospheric NO_x and no noticeable effect on ozone (Stolarski, Cicerone, and Nagy, 1973; Whitten and Turco, 1974*b*; Potter, 1978). It has been proposed that a large fleet of "super" shuttles could be used to carry men and materials into near space to build large structures for solar energy collection (DOE, 1980). A study of this problem has shown that the NO_x produced by ~400 super-shuttle launches and re-entries per year would deplete ozone by less than 1% (Whitten et al., 1982).

Nitrous Oxide (N_2O). Atmospheric nitrous oxide has its major source in bacterial denitrification processes (McElroy et al., 1976, 1977; Cicerone et al., 1978; Hahn, 1979) with secondary sources in combustion and lightning (Weiss and Craig, 1976; Pierotti and Rasmussen, 1976; Levine et al., 1979). Zipf and Prasad (1980, 1982) have also proposed an ionospheric source of N_2O. The human contribution is dominated by the use of manufactured fertilizers, whereby fixed nitrogen applied to the soil is partially reduced to N_2O. There is a debate at this time regarding the magnitude of the natural N_2O budget of the atmosphere and the importance of the anthropogenic component. If atmospheric N_2O were to build up significantly, its photodecomposition in the stratosphere would generate excess NO_x, as noted earlier (see Chapter 2).

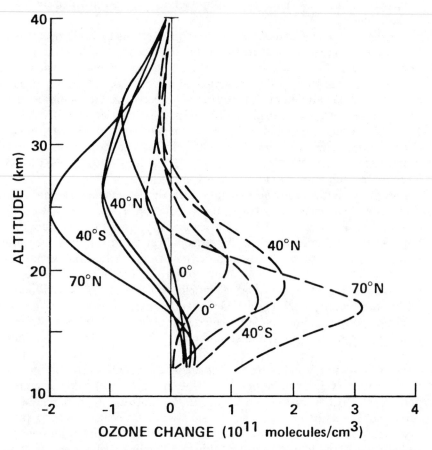

Figure 5-7. Steady-state ozone perturbations at several latitudes due to SST NO_x emissions at 20 km. The total injection is equivalent to 1.5×10 kg NO_2 per year. Results are shown for autumn in the Northern Hemisphere. The solid lines correspond to calculations using modern rate coefficients and the broken lines to older (circa 1979) rate coefficients. *(From Whitten et al., 1981, Implications of smaller concentrations of stratospheric OH: A two-dimensional model study of ozone perturbations, Atmos. Env.* **15.**)

For an assumed doubling of N_2O concentrations (from about 300 ppbv to 600 ppbv), current models predict a possible 10-16% global ozone depletion (steady-state). Figure 5-8 illustrates the changes in the ozone profile computed at different latitudes in an up-to-date 2-D model (Whitten et al., 1981, 1983). The effect on ozone is similar to that for massive SST NO_x injections, as reflected in the height dependence of the ozone depletion and the substantially larger ozone reductions at higher latitudes. Other multi-dimensional model studies of N_2O-ozone perturbations are discussed in the WMO/NASA (1982) report.

A slow increase of 0.2% per year has been detected (Weiss, 1981). Although the cause of the increase is unclear, it is obvious significant ozone perturbations could occur if man seriously interfered with the natural N_2O cycle.

Nuclear Explosions. Large nuclear detonations can inject massive quantities of NO_x into the stratosphere. It has been suggested that the atmospheric nuclear tests of the 1950s and 1960s perceptibly affected ozone concentrations (Foley and Ruderman, 1973; Johnston, Whitten, and Birks, 1973). Several attempts have been made to estimate the magnitude of the ozone change (Goldsmith et al., 1973; MacCracken and Chang, 1975; Bauer and Gilmore, 1975; Chang, Wuebbles, and Duewer,

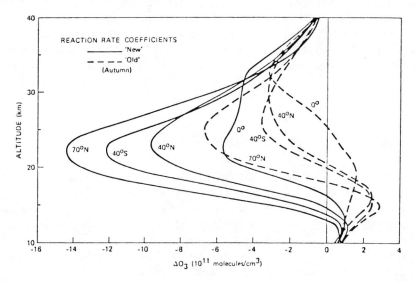

Figure 5-8. Predicted steady-state changes in stratospheric ozone concentrations at several latitudes when the atmospheric N_2O abundance is doubled. *(From Whitten et al., 1983, Revised predictions of the effect of increasing atmospheric N_2O and chlorofluoromethanes: A two-dimensional model study, Atmos. Env. **17**.)*

1979) and to identify the perturbation in the ozone record of that period (Johnston, Whitten, and Birks, 1973; Angell and Korshover, 1973, 1976). The data seem to limit the maximum possible ozone decrease to less than about 2%, while the models yield a decrease of about 2-3%. However, the uncertainties in the data and models are so large that firm conclusions cannot be drawn at this time.

Discussions of the effects of a full-scale nuclear war on stratospheric ozone must, of course, be strictly hypothetical. One would have difficulty imagining, for example, the primary destruction caused by 10,000 megatons (MT) of nuclear explosions (the present nuclear arsenals amount to about 13,000 MT). Roughly 50 million metric tons of NO could be generated by these explosions and deposited at high altitudes. Hampson (1974) first recognized that global depletion of the ozone layer by nuclear NO_x might lead to severe long-term worldwide environmental hazards to plant and animal species.

Photochemical model simulations of nuclear war effects on ozone have been carried out several times (e.g., Whitten, Borucki, and Turco, 1975a; MacCracken and Chang, 1975; National Academy of Sciences, 1975b; Crutzen and Birks, 1982). Results for a variety of war scenarios are shown in Figure 5-9. The striking feature is the large magnitude (up to 70% depletion) and duration (up to 5 years) of the perturbation. Recent 2-D simulations by Crutzen and Birks (1982) indicate that harmful uv-B radiation intensities at the ground could increase by a factor of ~8 at high latitudes (55°N) and by a factor of ~3 at low latitudes (~15°N), for roughly 1 year. Perturbations in the Southern Hemisphere are predicted to be much smaller because fewer nuclear bombs would be detonated there, and interhemispheric transfer of nuclear NO_x from the Northern Hemisphere should take a year or more (CIAP, 1975).

Halogens

Sources of anthropogenic halogens consist mainly of industrial emissions of volatile fluorocarbon refrigerants and various solvents, cleaning agents, fumigants, and related compounds (Cicerone, 1981). A unique source of HCl in the stratosphere is the space shuttle's solid-fueled booster engine. Chlorine is the most important halogen vis-à-vis stratospheric ozone change, with bromine and fluorine playing secondary roles. Iodine has not been detected in the stratosphere.

The natural background chlorine budget of the stratosphere (involving principally methyl chloride) amounts to ~2×10^7 Cl atoms/cm²-sec or ~1/5 million tons Cl per year. Periodic volcanic eruptions inject much smaller amounts (Cadle, 1975). The total "natural" stratospheric burden of chlorine is roughly 1/2 million tons, which accounts for less than about 5% of the background chemical destruction of ozone in the stratosphere.

Chlorofluoromethanes (CFMs). The chlorofluoromethanes (also, fluorocarbons or chlorofluorocarbons) are an extremely inert family of man-made compounds bearing the chemical formula $CH_xF_yCl_z$, with $x + y + z = 4$. No natural sources of fluorocarbons have been identified. The most widely used CFMs are $CFCl_3$ and CF_2Cl_2, or F-11 and F-12 respectively. Molina and Rowland (1974a) first suggested that the breakdown of chlorocarbons high in the stratosphere would release free chlorine and deplete ozone. Subsequently, a large number of modeling studies of past, present and future ozone reductions due to CFMs have been carried out (see Table 5-3). The historical record of CFM ozone depletion predictions is illustrated in Figure 5-10; the results are given for a steady-state chlorine increase of 1 ppbv due to F-11 and F-12 decomposition. At 1979 release rates, F-11 and F-12 could eventually add 5-6 ppbv of chlorine to the stratosphere.

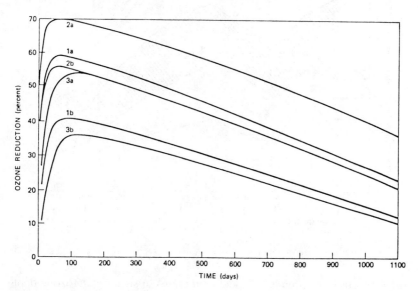

Figure 5-9. Total ozone depletions as a function of time after a series of large nuclear explosions, for several detonation scenarios: (1a, b) 5000 megatons (MT) divided equally between 1 and 5 MT explosions; (2a, b) 10,000 MT divided as in (1); (3a, b) 5000 MT divided equally between 1 and 3 MT explosions. The NO molecules produced by the explosions were assumed to be spread uniformly over (a) the latitude zone from 30°N to 70°N; (b) the Northern Hemisphere *(After Whitten, R. C., W. J. Borucki, and R. P. Turco, 1975a, Possible ozone depletions following nuclear explosions, Nature 257.)*

Cicerone, Walters, and Liu (1983) have determined that the latest chemical parameters in a one-dimensional model yield a highly nonlinear response of ozone to chlorine injection. This is illustrated in Figure 5-11. The change in ozone response is attributed mainly to a reduction in the simulated ambient OH concentrations; reduced OH levels moderate chlorine catalysis of ozone and enhance NO_x catalysis. These new results imply that the relative uncertainty in predicted absolute ozone depletions is very large, and that changes in total ozone due to chlorine pollutants will be more difficult to detect until significant contamination has occurred.

Figure 5-12 gives an example of the expected altitude dependence of the ozone reduction due to CFM consumption. Calculations corresponding to both old and new chemical reaction schemes are given. Two features are

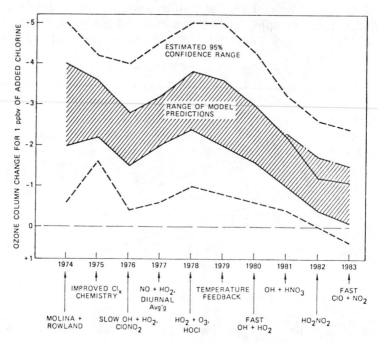

Figure 5-10. Historical trends in model predictions of steady-state ozone depletions due to 1 ppbv of chlorine added to the stratosphere as fluorocarbons. The range of depletion values calculated by different modelers is roughly bounded by the solid lines. A crude appraisal of the uncertainty range for 95% confidence in the predictions, held at the time of the calculations, is indicated by the dashed lines (Ehhalt et al., 1979). Key improvements in photochemical modeling techniques are shown in chronological order. The dash-dot line defines the average response for larger chlorine injections (i.e., $\%\Delta O_3/ppbv\text{-}Cl_x$) in cases where the ozone response is significantly nonlinear in Cl_x injection.

noteworthy. First, the local fractional ozone depletion is maximized at ~40 km. Thus, monitoring the topside of the ozone layer—perhaps by satellite—may provide the best means of detecting chlorine-induced ozone depletions while discriminating against NO_x and HO_x induced perturbations (WMO/NASA, 1982). Second, the lower region of ozone depletion (below 25 km), which is prominent in the older model calculations of Figure 5-12, is absent in the more recent predictions. In fact, ozone increases are now forecast. This explains most of the difference between the total ozone depletion estimates made between 1978 and 1982, as depicted in Figure 5-10. It also emphasizes the importance to model forecasts of ozone behavior at altitudes below 25 km, which is the

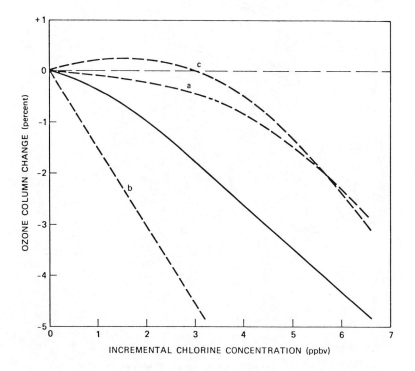

Figure 5-11. Calculated steady-state ozone column perturbations due to chlorine increases above background levels. The solid curve is based on 1982 chemistry with a 2 ppbv background Cl_x concentration. The dashed curves are taken from Cicerone, Walters, and Liu (1983). (a) a nominal updated chemistry set and ~1 ppbv of background Cl_x; (b) as in (a) but with reduced rates for the reactions of OH with HNO_3 and HO_2NO_2; (c) as in (a) but with a fast rate of formation of $ClONO_2$.

least-understood region of the stratosphere as regards photochemical and dynamical responses to human influences.

The possible time dependence of ozone reductions due to continuing CFM manufacture and release are illustrated in Figure 5-13. Some modern chemical schemes yield ozone curves about a factor of two lower than

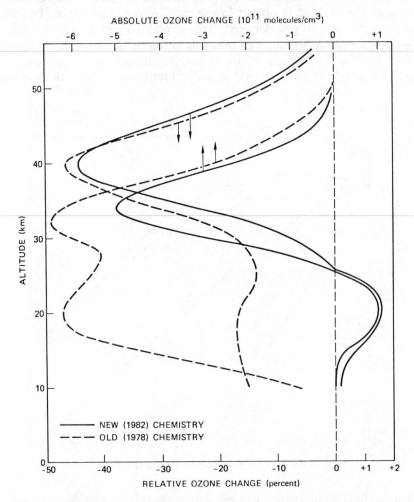

Figure 5-12. Predicted steady-state ozone reductions versus altitude for continuous fluorocarbon (CF_2Cl_2 and $CFCl_3$) emissions at 1979 production rates (Turco et al., 1981c). Both the relative and absolute ozone changes are given. Results are shown for 1978 and 1982 chemistry schemes. At steady-state, approximately 6 ppbv of chlorine (Cl_x) are produced by fluorocarbon decomposition. Note the change in the lower scale at the origin.

those shown (e.g., Cicerone, Walters, and Liu, 1983). Nonetheless, potential long-term ozone depletions are large, and they depend critically on man's decisions concerning the utilization of fluorocarbon compounds over the next few decades. Large ozone depletions might start to become evident early in the next century. Yet even if all fluorocarbon consumption were stopped in the year 2000, CFM-ozone depletions would persist through the twenty-first century.

Ozone trend curves such as those given in Figure 5-13 have been compared with time series of ozone measurements collected principally by the Dobson network (e.g., WMO/NASA, 1982). Such comparisons have not yet provided a meaningful test of the CFM-ozone theory or of model predictions in general. It may be concluded, however, that the observations are not inconsistent with the theory (and hence, with current CFM ozone decreases of ~0-1%). In fact, no direct evidence confirming the CFM-ozone depletion hypothesis has been uncovered, although the growing body of circumstantial evidence is impressive. Scientists have found the smoking gun, but the corpus delicti is still to be located.

The latitudinal and seasonal variations in CFM ozone perturbations have not been resolved. Calculations suggest a larger total ozone deple-

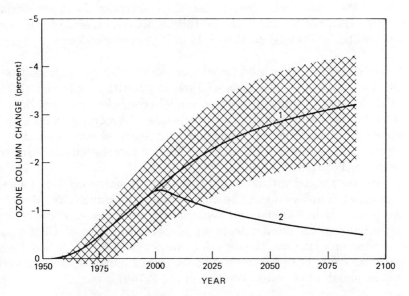

Figure 5-13. Typical model predictions of time-dependent ozone depletions for two fluorocarbon release scenarios beyond 1950. They are (1) F-11 and F-12 release at documented production rates until 1979, with constant release thereafter; (2) same as (1) but with no emissions beyond the year 2000. For curve (1), the cross-hatched area indicates the range of current model predictions.

tion at higher latitudes (by a factor of about two or less) and the possibility of a large seasonal variation (by a factor of about two or less), but only at high latitudes. Multidimensional model studies have been carried out by Brasseur and Bertin (1978), Derwent and Eggleton (1978), Vupputuri (1979), Pyle (1980), Borucki et al. (1980), Steed et al. (1982), and Whitten et al. (1981, 1983), among others.

Other Chlorocarbons. In addition to F-11 and F-12, the atmosphere holds a number of other manufactured chlorinated compounds including CCl_4 (carbon tetrachloride), C_2Cl_4 (tetrachloroethylene), CH_3CCl_3 (methyl chloroform), $CHClF_2$ (F-22), and CCl_2FCCl_2F (F-13). These compounds are discussed in the literature (Cicerone, 1975, 1981; Yung, McElroy, and Wofsy, 1975; Crutzen, Isaksen, and McAfee, 1978; Logan et al., 1978; WMO/NASA, 1982). Some of the chlorocarbon emissions leak into the stratosphere, where they decompose into free chlorine which catalytically attacks ozone. Based on current industrial emission rates, steady-state ozone depletions of 1-3% have been forecast if chlorocarbon usage continues at present levels.

The atmospheric lifetimes of these compounds range from ~50 years for CCl_4 to ~10 years for CH_3CCl_3 and $CHClF_2$ to much shorter times for some of the more complex molecules. Thus, except in the case of carbon tetrachloride, the related ozone perturbations are expected to be much smaller and less persistent than F-11 or F-12 perturbations.

Space Shuttle. The solid-fueled space shuttle booster engines deposit about 100 tons of HCl between ~15 and 45 km with each launch (Potter, 1978). Even so, 50 launches per year would account for less than about 1% of the total chlorine budget of the stratosphere. Accordingly, the steady-state ozone perturbation associated with space shuttle operations is expected to be about 0.1% or less. Detailed calculations confirm this estimate (Whitten et al., 1981).

Two-dimensional simulations of space shuttle emissions do not show a significant corridor effect at latitudes around the launch site. Moreover, the predicted latitudinal and seasonal dependences of shuttle ozone perturbations are similar in many respects to those of CFM ozone perturbations. However, the variations are not yet well established. Recovery of the ozone layer following the cessation of shuttle operations would require about 4 to 6 years (Whitten et al., 1975*b*).

Bromine. The stratospheric cycle of brominated compounds has not been clearly defined. Wofsy, McElroy, and Yung (1975) suggested that trace amounts of bromine might efficiently catalyze ozone destruction (Chapter 1). Although the importance of bromine has been argued in recent years (e.g., see Hudson and Reed, 1979; WMO/NASA, 1982), the amounts of gaseous bromine observed in the atmosphere are generally <10 pptv (parts per trillion by volume) (Berg et al., 1980; 1982). Even so, Yung et al. (1980) have calculated a possible 2% steady-state total ozone reduction from only 20 pptv of stratospheric bromine. It follows that bromine is a substance to be carefully monitored.

Fluorine. In the stratosphere, fluorine is primarily of interest as a tracer of fluorocarbon decomposition (Rowland and Molina, 1975). Because fluorine readily forms stable, unreactive compounds such as HF, projected future levels of fluorine are not expected to have any direct influence on ozone concentrations through photochemical activity. On the other hand, many fluorocarbon gases have strong infrared absorption bands, which might produce indirect effects on ozone if atmospheric temperatures and wind fields—to which ozone is sensitive—are inadvertently changed (see the discussion for CO_2 later in this chapter). In this regard, Cicerone (1979) pointed out that one anthropogenic fluorine gas, CF_4, has an atmospheric lifetime exceeding 10,000 years, and thus could accumulate over many centuries.

Water Vapor and Methane

Injections of water vapor, the precursor of free hydrogen radicals (HO_x), can lead to ozone perturbations. The principal anthropogenic source of water vapor is the exhaust of high-flying aircraft and rockets. Other pollutants that modify the temperature structure of the troposphere may also affect the tropopause "cold-trap" valve—which controls the natural circulation of water vapor through the stratosphere—and thus alter background stratospheric H_2O concentrations (Brewer, 1949; Liu et al., 1976*b*).

The water vapor budget of the stratosphere is uncertain, but methane decomposition alone can account for 60 million tons per year, and this may represent only 10% of the total. The stratosphere holds roughly 1,000 million tons of water vapor (Harries, 1976). Accordingly, massive injections of H_2O are required to produce noticeable perturbations.

Aircraft. All aircraft engines emit H_2O as a byproduct of fuel combustion. A large future fleet of SSTs might deposit hundreds of millions of tons of H_2O in the stratosphere. Figure 5-6 shows predictions of the effect of massive SST water vapor injections on total ozone. The impact is fairly small compared to that of NO_x, even assuming a low NO_2 emission index of ~6 g per kilogram of fuel burned, which corresponds to future engine technology goals (the H_2O emission index is essentially fixed at ~1300 g/kg by stoichiometric considerations).

Ozone perturbations caused by SST-H_2O injections would dissipate in about the same time as perturbations caused by SST-NO_x injections, that is, in about 2-4 years.

Rockets. Normal levels of rocket activity would not disturb the natural water cycle of the stratosphere. However, it has been proposed that "super" shuttles, or heavy lift launch vehicles (HLLVs), could be used in the future to construct a system of huge solar energy collectors in space (DOE, 1980). The project would require ~400 HLLV launches per year, each launch depositing ~3800 tons of H_2O in the stratosphere and mesosphere. The predicted decrease in ozone corresponding to these water injections, however, is only ~0.05% (Whitten et al., 1982).

Because of the high altitudes of injection of HLLV water vapor, it would require a somewhat longer time to diffuse out of the stratosphere than SST H_2O (probably ~5-7 years).

Methane. Methane affects the chemistry of the lower stratosphere, and thus the concentration of ozone, by interacting with the hydrogen and chlorine cycles and initiating smog reactions. Initial studies of tropospheric methane variations related to anthropogenic CO emissions were made by Chameides, Liu, and Cicerone (1977) and Sze (1977). Observations indicate that atmospheric CH_4 concentrations are slowly increasing (Graedel and McRae, 1980; Fraser et al., 1981; Blake et al., 1982). Simulations carried out by Owens et al. (1982) suggest that a doubling of methane concentrations might lead to a 3% ozone increase, if all other factors remain constant.

Nonmethane hydrocarbons (Chameides and Cicerone, 1978) and molecular hydrogen (Penner, McElroy, and Wofsy, 1977) can produce chemical effects similar to those of methane in the atmosphere. However, very few studies of such species have been carried out (e.g., Aikin et al., 1982).

Carbon Dioxide

The carbon dioxide burden of the atmosphere appears to be increasing due to massive consumption of fossil fuels (National Research Council, 1979; CEQ, 1981). Carbon dioxide cools the stratosphere by emitting infrared radiation to space. An increase in CO_2 concentrations can therefore decrease stratospheric air temperatures. At the same time, CO_2 creates a tropospheric greenhouse effect and warms the lower atmosphere. The rate coefficients of many of the chemical reactions that control ozone are sensitive to temperature. In particular, as the temperature decreases, the ozone concentration increases. Because the effect is indirect, and requires both an accurate prediction of small temperature changes and a detailed knowledge of rate coefficient temperature dependences, CO_2-induced ozone perturbations are subject to additional uncertainty. Moreover, because temperature fields and dynamical motions are closely coupled in the atmosphere, induced variations in ozone transport rates must also be taken into account.

For a doubling of CO_2, recent estimates place the steady-state ozone column increase at $\sim 3\text{-}9\%$ (Groves, Mattingly, and Tuck, 1978; Haigh and Pyle, 1979; Fels et al., 1980; Callis and Natarajan, 1981). The maximum temperature decrease, near 40 km altitude, is about 10 K. The persistence of the CO_2 effect is related to the geological time-scale for assimilation and sequestering of atmospheric carbon in the oceans and on the continents. The time-scale is uncertain, but it probably falls in the range of 100-1000 years (Siegenthaler and Oeschger, 1978).

Summary

The most critical potential anthropogenic influences on the stratospheric ozone layer are summarized in Table 5-5. Also indicated are the persistence of effects following termination of pollution, and the mechanism of ozone recovery.

MULTIPLE PERTURBATIONS

Up to this point, estimates of ozone change have focused on single perturbing agents acting alone. This procedure defines the relative importance and potential effects of individual pollutants. However, simultaneous perturbations must also be considered for two reasons. First, the earth's ozone layer is under the concerted influence of a number of chemical and physical agents. Second, interactions between these agents can lead to a net change in ozone that is quite different from the sum of the expected individual changes. For example, simultaneous NO_x and Cl_x injections are less effective, overall, in reducing ozone than are the separate injections. The photochemical aspects of these complex multicontaminant interactions are reviewed in Chapter 2.

Consideration of simultaneous perturbations represents an improvement when model simulations are to be compared with observational data on ozone variations. However, when ozone change must be projected into the future, multiple-pollutant scenarios create additional difficulties.

Table 5-5. Anthropogenic Perturbations of Stratospheric Ozone

Perturbing Agent	Ozone Change (%)	Recovery Time[a] (years)	Mode of Pollutant Removal
SST-NO$_x$ 1 × 10^9 kg NO$_2$/yr @ 20 km	−4 to −8[b]	2-4	Transport of injected NO$_x$ to the troposphere
Subsonic Aircraft 2 × 10^9 kg NO$_2$/yr @ 12 km	−1 to +2[b]	1-3	Dispersion of injected NO$_x$ from the tropopause region
N$_2$O Doubling	−10 to −16[b]	~100?	Photochemical depletion of N$_2$O from the tropospheric reservoir
Nuclear war 10,000 Megatons, or 10^{36} NO molecules	−50 to −70[c]	3-6	Transport of injected NO$_x$ to the troposphere
CFMs F-11 and F-12 ~5 ppbv Cl$_x$	−3 to −10[b]	50-150	Fluorocarbon photolysis in the middle stratosphere
CH$_x$Cl$_y$ Chlorocarbons ~2 ppbv Cl$_x$	−1 to −3[b]	1-15	Decomposition of the tropospheric chlorocarbon reservoir by reaction with OH
CO$_2$ doubling	+3 to +9[b]	100-1000	Recycling of excess CO$_2$ into the biosphere oceans and geological deposits
CH$_4$ doubling	+1 to +5[b]	~10	Decomposition of excess tropospheric CH$_4$ by reaction with OH

[a] Following abrupt termination of pollutant emissions.
[b] Steady-state change.
[c] Maximum change, occurring several months after injection.

Table 5-6. Relative Effectiveness of Simultaneous Ozone Perturbations

Perturbation	Relative Efficiency[a] for Ozone Change
CFMs + 2 × N_2O	~0.8
CFMs + 2 × CO_2	~1.1–2.0
CFMs + 2 × CH_4	~0.5
CFMs + 20 pptv Bromine	~1.1–1.2
SST NO_x + H_2O (20 km)	~0.8
2 × N_2O + 2 × CH_4	~0.7

[a] Defined as the ratio of the compound perturbation to the sum of the individual perturbations, at steady-state.

For one thing, they increase the number of parameters required to specify the contaminants and their emission scenarios. Thus, a larger number of calculations must be made to define the range of possible outcomes. The uncertainty in the predictions may also be greater, in the sense that concurrent changes in a large number of emission parameters under human (societal, economic) control are exceedingly difficult to forecast.

Multiple-pollutant ozone simulations have only recently been emphasized. The status of such calculations is reviewed in the WMO/NASA Report (1982). Some estimates of the degree of interaction that occurs between key ozone-active chemical agents are summarized in Table 5-6 (Yung et al., 1980; Callis and Natarajan, 1981; WMO/NASA, 1982; Wuebbles, Luther, and Penner, 1983). The estimates are very crude and should only be used as rough indications of ozone change, particularly when the effect of a CO_2 increase on the magnitude of another ozone perturbation is considered. For example, the CFM + 2 × CO_2 case in Table 5-6 shows an apparent amplification effect, because the ozone changes due to CFM and CO_2 increases are nearly equal in magnitude but have opposite signs. Hence, the net impact on ozone is small and difficult to quantify accurately (even the sign of the net change is in some doubt).

CONCLUSIONS

Ozone is obviously sensitive to a variety of natural events and human activities. Computer forecasts of ozone change suggest that existing or accelerated rates of anthropogenic pollution will eventually cause serious environmental damage on a global scale. The large natural variability of the atmosphere has so far concealed any evidence of human influence on ozone. Also, direct observational confirmation of ozone perturbation theories is lacking, which has led to uncertainty and indecision in formulating pollution control strategies.

Nevertheless, the future looks promising. There is a growing solid base of fundamental photochemical data which can be used to make reliable model predictions. Atmospheric measurements are improving in precision, geographical coverage, and time resolution, and are defining in greater detail the composition and state of the present atmosphere. Significantly, there is a growing belief that the environment must be protected from excessive contamination until scientists can develop tools capable of forecasting ozone change accurately. The future challenge will be to synthesize existing complex physical theories with extensive observational data to create realistic and functional representations of the earth's stratosphere.

REFERENCES

Agashe, V. V., and S. M. Rathi, 1982, Changes in the concentration of mesospheric ozone during the total solar eclipse, *Planet. Space Sci.* **30**:507-513.

Aikin, A. C., S. Chandra, and T. P. Stecher, 1980, Supernovae effects on the terrestrial atmosphere, *Planet. Space Sci.* **28**:639-644.

Aikin, A. C., J. R. Herman, E. J. Maier, and C. J. McQuillan, 1982, Atmospheric chemistry of ethane and ethylene, *J. Geophys. Res.* **87**:3105-3118.

Aimedieu, P., P. Rigaud, and J. Barat, 1981, The sunrise ozone depletion problem of the upper stratosphere, *Geophys. Res. Lett.* **8**:787-789.

Alyea, F. N., D. M. Cunnold, and R. G. Prinn, 1975, Stratospheric ozone destruction by aircraft-induced nitrogen oxides, *Science* **188**:117-121.

Angell, J. K., and J. Korshover, 1973, Quasi-biennial and long-term fluctuations in total ozone, *Mon. Weather Rev.* **101**:426-443.

Angell, J. K., and J. Korshover, 1976, Global analysis of recent total ozone fluctuations, *Mon. Weather Rev.* **104**:63-75.

Angell, J. K., and J. Korshover, 1978, Comparison of stratospheric trends in temperature, ozone and water vapor in north temperate latitudes, *J. Appl. Meteorol.* **17**:1397-1401.

Bates, D. R., and P. B. Hays, 1967, Atmospheric nitrous oxide, *Planet. Space Sci.* **15**:189-197.

Bates, D. R., and M. Nicolet, 1950, The photochemistry of atmospheric water vapor, *J. Geophys. Res.* **55**:301-327.

Bauer, E., 1979, A catalog of perturbing influences on stratospheric ozone, 1955-1975, *J. Geophys. Res.* **84**:6929-6940.

Bauer, E., and F. R. Gilmore, 1975, Effect of atmospheric nuclear explosions on total ozone, *Rev. Geophys. Space Phys.* **13**:451-458.

Berg, W. W., P. J. Crutzen, F. E. Grahek, S. N. Gitlin, and W. A. Sedlacek, 1980, First measurements of total chlorine and bromine in the lower stratosphere, *Geophys. Res. Lett.* **7**:937-940.

Berg, W. W., F. E. Grahek, E. S. Gladney, and W. A. Sedlacek, 1982, The global bromine budget: Recent measurements and theories, paper presented at the Second Symposium on the Composition of the Nonurban Troposphere (preprint available from American Meteorological Society, Boston, Mass.).

Berkner, L. W., and L. L. Marshall, 1967, The rise of oxygen in the Earth's atmosphere with notes on the Martian atmosphere, *Adv. Geophys.* **12**:309-331.

Blackshear, W. T., and R. H. Tolson, 1978, High correlations between variations in monthly averages of solar activity and total atmospheric ozone, *Geophys. Res. Lett.* **5**:921-924.

Blake, A. J., and J. H. Carver, 1977, The evolutionary role of atmospheric ozone, *J. Atmos. Sci.* **34**:720-728.

Blake, D. R., E. W. Mayer, S. C. Tyler, Y. Makide, D. C. Montague, and F. S. Rowland, 1982, Global increase in atmospheric methane concentrations between 1978 and 1980, *Geophys. Res. Lett.* **9**:477-480.

Borucki, W. J., R. C. Whitten, V. R. Watson, H. T. Woodward, C. A. Riegel, L. A. Capone, and T. Becker, 1976, Model predictions of latitude-dependent ozone depletion due to supersonic transport operations, *AIAA J.* **14**:1738-1745.

Borucki, W. J., R. C. Whitten, H. T. Woodward, L. A. Capone, C. A. Riegel, and S. Gaines, 1980, Stratospheric ozone decrease due to chlorofluoromethane photolysis: Predictions of latitude dependence, *J. Atmos. Sci.* **37**:686-697.

Brasseur, G., and M. Bertin, 1978, The action of chlorine on the ozone layer as given by a zonally averaged two-dimensional model, *Pure Appl. Geophys.* **117**:436-447.

Brewer, A. W., 1949, Evidence for a world circulation provided by the measurements of helium and water vapour distribution in the stratosphere, *R. Meteorol. Soc. Q. J.* **75**:351-363.

Broderick, A., 1977, Stratospheric effects from aviation, AIAA 13th Propulsion Conference, Orlando, Fla., July 11-13, 1977, Paper 77-799.

Brueckner, G. E., 1981, The variability of the Sun's ultraviolet radiation, *Planet. Aeron. Astron. Adv. Space Res.* **7**:101-116.

Cadle, R. D., 1975, Volcanic emissions of halides and sulfur compounds to the troposphere and stratosphere, *J. Geophys. Res.* **80**:1650-1652.

Cadle, R. D., 1980, Some effects of the emissions of explosive volcanoes on the stratosphere, *J. Geophys. Res.* **85**:4495-4498.

Callis, L. B., and M. Natarajan, 1981, Atmospheric carbon dioxide and chlorofluoromethanes: Combined effects on stratospheric ozone, temperature, and surface temperature, *Geophys. Res. Lett.* **8**:587-590.

Callis, L. B., and J. E. Nealy, 1978, Solar UV variability and its effect on stratospheric thermal structure and trace constituents, *Geophys. Res. Lett.* **5**:249-252.

CEQ, 1981, *Global Energy Futures and the Carbon Dioxide Problem,* Council on Environmental Quality, U.S. Government Printing Office, Washington, D.C.

Chakrabarty, D. K., and P. Chakrabarty, 1982, The evolution of ozone with changing solar activity, *Geophys. Res. Lett.* **9**:76-78.

Chameides, W. L., and R. J. Cicerone, 1978, Effects of nonmethane hydrocarbons in the atmosphere, *J. Geophys. Res.* **83**:947-952.

Chameides, W. L., S. C. Liu, and R. J. Cicerone, 1977, Possible variations in atmospheric methane, *J. Geophys. Res.* **82**:1795-1798.

Chang, J. S., D. J. Wuebbles, and W. H. Duewer, 1979, The atmospheric nuclear tests of the 1950's and 1960's: A possible test of ozone depletion theories, *J. Geophys. Res.* **84**:1755-1765.

CIAP, 1975, *The Stratosphere Perturbed by Propulsion Effluents*, CIAP Monograph III, DOT-TST-75-53, Department of Transportation, Washington, D.C.

Cicerone, R. J., 1975, Comment on "Volcanic emissions of halides and sulfur compounds to the troposphere and stratosphere" by R. D. Cadle, *J. Geophys. Res.* **80**:3911-3912.

Cicerone, R. J., 1979, Atmospheric carbon tetrafluoride: A nearly inert gas, *Science* **206**:59-61.

Cicerone, R. J., 1981, Halogens in the atmosphere, *Rev. Geophys. Space Phys.* **19**:123-139.

Cicerone, R. J., R. S. Stolarski, and S. Walters, 1974, Stratospheric ozone destruction by man-made chlorofluoromethanes, *Science* **185**:1165-1167.

Cicerone, R. J., S. Walters, and S. C. Liu, 1983, Nonlinear response of stratospheric ozone column to chlorine injections *J. Geophys. Res.* **88**:3647-3661.

Cicerone, R. J., S. Walters, and R. S. Stolarski, 1975, Chlorine compounds and stratospheric ozone, *Science* **188**:378-379.

Cicerone, R. J., J. D. Shetter, D. H. Stedman, T. J. Kelly, and S. C. Liu, 1978, Atmospheric N_2O: Measurements to determine its sources, sinks, and variations, *J. Geophys. Res.* **83**:3042-3050.

COMESA, 1975, *The Report of the Committee on Meteorological Effects of Stratospheric Aircraft (1972-1975)*, U. K. Meteorological Office, Bracknell.

Cook, J. W., G. E. Brueckner, and M. E. VanHoosier, 1980, Variability of the solar flux in the far ultraviolet 1175-2100 Å, *J. Geophys. Res.* **85**:2257-2268.

Crutzen, P. J., 1970, The influence of nitrogen oxides on the atmospheric ozone content, *R. Meteorol. Soc. Q. J.* **96**:320-325.

Crutzen, P. J., 1971, Ozone production rates in an oxygen-hydrogen-nitrogen oxide atmosphere, *J. Geophys. Res.* **76**:7311-7327.

Crutzen, P. J., 1972, SST's: A threat to the earth's ozone shield, *Ambio* **1**:41-51.

Crutzen, P. J., 1973, Gas-phase nitrogen and methane chemistry in the atmosphere, in *Physics and Chemistry of the Upper Atmosphere*, B. M. McCormac, ed., Reidel, Dordrecht, Holland, 110-124.

Crutzen, P. J., 1974a, Estimates of possible variations in total ozone due to natural causes and human activities, *Ambio* **3**:201-210.

Crutzen, P. J., 1974b, Estimates of possible future ozone reductions from continued use of fluoro-chloro-methanes (CF_2Cl_2, $CFCl_3$), *Geophys. Res. Lett.* **1**:205-208.

Crutzen, P. J., 1976, Upper limits on atmospheric ozone reductions following increased application of fixed nitrogen to the soil, *Geophys. Res. Lett.* **3**:169-172.

Crutzen, P. J., and J. W. Birks, 1982, The atmosphere after a nuclear war: Twilight at noon, *Ambio* **11**:114-125.

Crutzen, P. J., I. S. A. Isaksen, and J. R. McAfee, 1978, The impact of the chlorocarbon industry on the ozone layer, *J. Geophys. Res.* **83**:345-363.

Cunnold, D., F. Alyea, N. Phillip, and R. Prinn, 1975, A three-dimensional dynamical-chemical model of atmospheric ozone, *J. Atmos. Sci.* **32**:170-194.

Cunnold, D. M., F. N. Alyea, and R. G. Prinn, 1980, Preliminary calculations concerning the maintenance of the zonal mean ozone distribution in the northern hemisphere, *Pure Appl. Geophys.* **118**:329-354.

Derwent, R. G., and E. J. Eggleton, 1978, Halocarbon lifetimes and concentration distributions calculated using a two-dimensional tropospheric model, *Atmos. Env.* 12:1261-1269.

Derwent, R. G., and E. J. Eggleton, 1981, Two-dimensional model studies of methyl chloroform in the troposphere, *R. Meteorol. Soc. Q. J.* 107:231-242.

DOE, 1980, *Proceedings of the Workshop on the Modification of the Upper Atmosphere by Satellite Power System Propulsion Effluents*, Report CONF-7906180, U.S. Department of Energy, Washington, D.C.

Dotto, L., and H. Schiff, 1978, *The Ozone War*, Doubleday, Garden City, New York.

Dütsch, H. U., 1979, The search for solar-cycle ozone relationships, *J. Atmos. Terr. Phys.* 41:771-785.

Dütsch, H. U., and W. Braun, 1980, Daily ozone soundings during two winter months including a sudden stratospheric warming, *Geophys. Res. Lett.* 7:785-788.

Ehhalt, D. H., J. S. Chang, and D. M. Butler, 1979, The probability distribution of the predicted CFM-induced ozone depletion, *J. Geophys. Res.* 84:7889-7894.

Fabian, P., J. A. Pyle, and R. J. Wells, 1979, The August 1972 solar proton event and the atmospheric ozone layer, *Nature* 277:458-460.

Fabian, P., J. A. Pyle, and R. J. Wells, 1982, Diurnal variations of minor constituents in the stratosphere modeled as a function of latitude and season, *J. Geophys. Res.* 87:4981-5000.

Fels, S. B., J. D. Mahlman, M. D. Schwarzkopf, and R. W. Sinclair, 1980, Stratospheric sensitivity to perturbations in ozone and carbon dioxide: Radiative and dynamical response, *J. Atmos. Sci.* 37:2265-2297.

Fiocco, G., A. Mugnai, and W. Forlizzi, 1978, Effects of radiation scattered by aerosols on the photodissociation of ozone, *J. Atmos. Terr. Phys.* 40:949-961.

Foley, H. M., and M. A. Ruderman, 1973, Stratospheric NO production from past nuclear explosions, *J. Geophys. Res.* 78:4441-4450.

Fraser, P. J., M. A. K. Khalil, R. A. Rasmussen, and A. J. Crawford, 1981, Trends of atmospheric methane in the southern hemisphere, *Geophys. Res. Lett.* 8:1063-1066.

Frederick, J. E., R. B. Abrams, R. Dasgupta, and B. Guenther, 1980, An observed annual cycle in tropical upper stratospheric and mesospheric ozone, *Geophys. Res. Lett.* 7:713-716.

Gandrud, B. W., and A. L. Lazrus, 1981, Filter measurements of stratospheric sulfate and chloride in the eruption plume of Mount St. Helens, *Science* 211:826-827.

Gandrud, B. W., and A. L. Lazrus, 1984, Measurements of stratospheric sulfate and acidic chloride attributed to El Chichon, *Science,* in press.

Ghazi, A., 1974, Nimbus 4 observations of changes in total ozone and stratospheric temperatures during a sudden warming, *J. Atmos. Sci.* 31:2197-2206.

Goldsmith, P., A. F. Tuck, J. S. Foot, E. L. Simmons, and R. L. Newson, 1973, Nitrogen oxides, nuclear weapon testing, Concorde and stratospheric ozone, *Nature* 244:545-551.

Graedel, T. E., and J. E. McRae, 1980, On the possible increase of the atmospheric methane and carbon monoxide concentrations during the last decade, *Geophys. Res. Lett.* 7:977-979.

Grobecker, A. J., S. C. Coroniti, and R. H. Cannon, Jr., 1974, *The Effects of*

Stratospheric Pollution by Aircraft, DOT-TST-75-50, Dept. of Transportation, Washington, D.C.

Groves, K. S., and A. F. Tuck, 1980, Stratospheric O_3-CO_2 coupling in a photochemical-radiative column model. II: With chlorine chemistry, *R. Meteorol. Soc. Q. J.* **106**:141-157.

Groves, K. S., S. R. Mattingly, and A. F. Tuck, 1978, Increased atmospheric carbon dioxide and stratospheric ozone, *Nature* **273**:711-715.

Hahn, J., 1979, The cycle of atmospheric nitrous oxide, *R. Soc. (London) Philos. Trans.* **A290**:495-504.

Haigh, J. D., and J. A. Pyle, 1979, A two-dimensional calculation including atmospheric carbon dioxide and stratospheric ozone, *Nature* **279**:222-224.

Hampson, J., 1974, Photochemical war on the atmosphere, *Nature* **250**:189-191.

Harries, J. E., 1976, The distribution of water vapor in the stratosphere, *Rev. Geophys. Space Phys.* **14**:565-575.

Harrison, H., 1970, Stratospheric ozone with added water vapor: Influence of high-altitude aircraft, *Science* **170**:734-736.

Harwood, R. S., and J. A. Pyle, 1977, Studies of the ozone budget using a zonal mean circulation model and linearized photochemistry, *R. Meteorol. Soc. Q. J.* **103**:319-343.

Heath, D. F., 1973, Space observations of the variability of solar irradiance in the near and far ultraviolet, *J. Geophys. Res.* **78**:2779-2792.

Heath, D. F., A. J. Krueger, and P. J. Crutzen, 1977, Solar proton event: Influence on stratospheric ozone, *Science* **197**:886-889.

Herman, J. R., 1979, The response of stratospheric constituents to a solar eclipse, sunrise, and sunset, *J. Geophys. Res.* **84**:3701-3710.

Hidalgo, H., and P. J. Crutzen, 1977, The tropospheric and stratospheric composition perturbed by NO_x emissions of high altitude aircraft, *J. Geophys. Res.* **82**:5833-5866.

Hilsenrath, E., 1980, Rocket observations of the vertical distribution of ozone in the polar night and during a mid-winter stratospheric warming, *Geophys. Res. Lett.* **7**:581-584.

Hilsenrath, E., D. F. Heath, and B. M. Schlesinger, 1979, Seasonal and interannual variations in total ozone revealed by the Nimbus 4 backscattered ultraviolet experiment, *J. Geophys. Res.* **84**:6969-6979.

Hofman, D. J., and J. M. Rosen, 1983, Stratospheric sulfuric acid fraction and mass estimate for the 1982 volcanic eruption of El Chichon, *Geophys. Res. Lett.* **4**:313-316.

Holton, J. R., 1979, *An Introduction to Dynamic Meteorology*, Academic Press, New York.

Hudson, R. D., ed., 1977, *Chlorofluoromethanes and the Stratosphere*, Ref. Publ. 1010, National Aeronautics and Space Administration, Washington, D.C.

Hudson, R. D., and E. I. Reed, eds., 1979, *The Stratosphere: Present and Future*, Ref. Publ. 1049, National Aeronautics and Space Administration, Washington, D.C.

Johnston, D. A., 1980, Volcanic contribution of chlorine to the stratosphere: More significant to ozone than previously estimated?, *Science* **209**:491-492.

Johnston, H., 1971, Reduction of stratospheric ozone by nitrogen oxide catalysts from supersonic transport exhaust, *Science* **173**:517-522.

Johnston, H. S., 1977, Analysis of the independent variables in the perturbation of stratospheric ozone by nitrogen fertilizers, *J. Geophys. Res.* 82:1767-1772.

Johnston, H. S., and J. Podolske, 1978, Interpretations of stratospheric photochemistry, *Rev. Geophys. Space Phys.* 16:491-519.

Johnston, H., G. Whitten, and J. Birks, 1973, Effect of nuclear explosions on stratospheric nitric oxide and ozone, *J. Geophys. Res.* 78:6107-6135.

Kasting, J. F., and T. M. Donahue, 1980, The evolution of atmospheric ozone, *J. Geophys. Res.* 85:3255-3263.

Keating, G. M., 1978, Relation between monthly variations of global ozone and solar activity, *Nature* 274:873-874.

Keating, G. M., L. R. Lake, J. Y. Nicholson III, and M. Natarajan, 1981, Global ozone long-term trends from satellite measurements and the response to solar activity variations, *J. Geophys. Res.* 86:9873-9880.

Krueger, A. J., 1983, Sighting of El Chichon sulfur dioxide clouds with the Nimbus 7 total ozone mapping spectrometer, *Science* 220:1377-1379.

Lazrus, A. L., R. D. Cadle, B. W. Gandrud, J. P. Greenberg, B. J. Huebert, and W. I. Rose, Jr., 1979, Sulfur and halogen chemistry of the stratosphere and of volcanic eruption plumes, *J. Geophys. Res.* 84:7869-7875.

Lean, J. L., 1982, Observation of the diurnal variation of atmospheric ozone, *J. Geophys. Res.* 87:4973-4980.

Levine, J. S., R. E. Hughes, W. L. Chameides, and W. E. Howell, 1979, N_2O and CO production by electric discharge: Atmospheric implications, *Geophys. Res. Lett.* 6:557-559.

Lewis, J. S., G. H. Watkins, H. Hartman, and R. G. Prinn, 1982, *Chemical Consequences of Major Impact Events on Earth,* Geological Society of America Special Paper 190, L. T. Silver and P. H. Schultz, eds., 215-221.

Liu, S. C., R. J. Cicerone, T. M. Donahue, and W. L. Chameides, 1976a, Limitation of fertilizer induced ozone reduction by the long lifetime of the reservoir of fixed nitrogen, *Geophys. Res. Lett.* 3:157-160.

Liu, S. C., T. M. Donahue, R. J. Cicerone, and W. L. Chameides, 1976b, Effect of water vapor on the destruction of ozone in the stratosphere perturbed by Cl_x or NO_x pollutants, *J. Geophys. Res.* 81:3111-3118.

Logan, J. A., M. J. Prather, S. C. Wofsy, and M. B. McElroy, 1978, Atmospheric chemistry: Response to human influence, *R. Soc. London Philos. Trans., ser. A,* 290:187-234.

London, J., and J. Angell, 1982, The observed distribution of ozone and its variations, *Stratospheric Ozone and Man, Vol. 1, Stratospheric Ozone,* F. A. Bower and R. B. Ward, eds., CRC press, Boca Raton, Fla.

London, J., and C. A. Reber, 1979, Solar activity and total atmospheric ozone, *Geophys. Res. Lett.* 6:869-872.

London, J., J. E. Frederick, and G. P. Anderson, 1977, Satellite observations of the global distribution of stratospheric ozone, *J. Geophys. Res.* 82:2543-2556.

Luther, F. M., D. Wuebbles, and J. Chang, 1977, Temperature feedback in a stratospheric model, *J. Geophys. Res.* 82:4935-4942.

McConnell, J. C., and H. I. Schiff, 1978, Methyl chloroform: Impact on stratospheric ozone, *Science* 199:174-176.

MacCracken, M. C., and J. S. Chang, 1975, A preliminary study of the potential chemical and climatic effects of atmospheric nuclear explosions, Univ. of California Livermore Laboratory, Report UCRL-51653, 77 pp.

McElroy, M. B., and J. C. McConnell, 1971, Nitrous oxide: A natural source of stratospheric NO, *J. Atmos. Sci.* **28**:1095-1098.

McElroy, M. B., S. C. Wofsy, and Y. L. Yung, 1977, The nitrogen cycle: Perturbations due to man and their impact on atmospheric N_2O and O_3, *R. Soc. London Philos. Trans., ser. B,* **277**:159-181.

McElroy, M. B., S. C. Wofsy, J. E. Penner, and J. C. McConnell, 1974, Atmospheric ozone: Possible impact of stratospheric aviation, *J. Atmos. Sci.* **31**:287-303.

McElroy, M. B., J. W. Elkins, S. C. Wofsy, and Y. L. Yung, 1976, Sources and sinks for atmospheric N_2O, *Rev. Geophys. Space Phys.* **14**:143-150.

McPeters, R. D., C. H. Jackman, and E. G. Stassinopoulos, 1981, Observations of ozone depletion associated with solar proton events, *J. Geophys. Res.* **86**:12,071-12,081.

Mankin, W. G., and M. T. Coffey, 1984, Increased stratospheric hydrogen chloride in the El Chicón cloud, *Science* **226**:170-172.

Miller, C., D. L. Filkin, A. J. Owens, J. M. Steed, and J. P. Jesson, 1981, A two-dimensional model of stratospheric chemistry and transport, *J. Geophys. Res.* **86**:12,039-12,065.

Molina, M. J., and F. S. Rowland, 1974a, Stratospheric sink for chlorofluoromethanes: Chlorine atom-catalysed destruction of ozone, *Nature* **249**:810-812.

Molina, M. J., and F. S. Rowland, 1974b, Predicted present stratospheric abundances of chlorine species from photodissociation of carbon tetrachloride, *Geophys. Res. Lett.* **1**:309-312.

National Academy of Sciences, 1975a, *Environmental Impact of Stratospheric Flight,* Washington, D.C.

National Academy of Sciences, 1975b, *Long-term Worldwide Effects of Multiple Nuclear-Weapons Detonations,* Washington, D.C.

National Academy of Sciences, 1976, *Halocarbons: Effects on Stratospheric Ozone,* Washington, D.C.

National Academy of Sciences, 1979, *Stratospheric Ozone Depletion By Halocarbons: Chemistry and Transport,* Washington, D.C.

National Research Council, 1979, *Carbon Dioxide and Climate, a Scientific Assessment,* National Academy of Sciences, Washington, D.C.

Nicolet, M., 1975, Stratospheric ozone: An introduction to its study, *Rev. Geophys. Space Phys.* **13**:593-636.

Owens, A. J., J. M. Steed, D. L. Filkin, C. Miller, and J. P. Jesson, 1982, The potential effects of increased methane on atmospheric ozone, *Geophys. Res. Lett.* **9**:1105-1108.

Park, C., 1978, Nitric oxide produced by the Tunguska meteor, *Acta Astronaut.* **5**:523-542.

Park, C., and G. P. Menees, 1978, Odd-nitrogen production by meteoroids, *J. Geophys. Res.* **83**:4029-4035.

Penner, J. E., and J. S. Chang, 1978, Possible variations in atmospheric ozone related to the eleven-year solar cycle, *Geophys. Res. Lett.* **5**:817-820.

Penner, J. E., and J. S. Chang, 1980, The relation between atmospheric trace species variabilities and solar uv variability, *J. Geophys. Res.* **85**:5523-5528.

Penner, J. E., M. B. McElroy, and S. C. Wofsy, 1977, Sources and sinks for atmospheric H_2: A current analysis with projections for the influence of anthropogenic activity, *Planet. Space Sci.* **25**:521-540.

Pierotti, D., and R. A. Rasmussen, 1976, Combustion as a source of nitrous oxide in the atmosphere, *Geophys. Res. Lett.* **3:**265-267.

Pollack, J. B., and Y. L. Yung, 1980, Origin and evolution of planetary atmospheres, *Ann. Rev. Earth Planet. Sci.* **8:**425-487.

Poppoff, I. G., R. C. Whitten, R. P. Turco, and L. A. Capone, 1978, *An Assessment of the Effect of Supersonic Aircraft Operations on the Stratospheric Ozone Content,* Ref. Publ. 1026, National Aeronautics and Space Administration, Washington, D.C.

Potter, A. E., 1978, *Revised Estimates for Ozone Reduction by Shuttle Operation,* NASA Tech. Mem. 58209, Lyndon B. Johnson Space Center.

Pyle, J. A., 1980, A calculation of the possible depletion of ozone by chlorofluorocarbons using a two-dimensional model, *Pure Appl. Geophys.* **118:**355-377.

Reagan, J. B., R. E. Meyerott, R. W. Nightingale, R. C. Gunton, R. G. Johnson, J. E. Evans, W. L. Imhof, D. F. Heath, and A. J. Krueger, 1981, Effects of the August 1972 solar particle events on stratospheric ozone, *J. Geophys. Res.* **86:**1473-1494.

Reber, C. A., and F. T. Huang, 1982, Total ozone-solar activity relationships, *J. Geophys. Res.* **87:**1313-1318.

Reid, G. C., J. R. McAfee, and P. J. Crutzen, 1978, Effects of intense stratospheric ionisation events, *Nature* **275:**489-492.

Rowland, F. S., and M. J. Molina, 1975, Chlorofluoromethanes in the environment, *Rev. Geophys. Space Phys.* **13:**1-35.

Ruderman, M. A., and J. W. Chamberlain, 1975, Origin of the sunspot modulation of ozone: Its implications for stratospheric NO injection, *Planet. Space Sci.* **23:**247-268.

Ruderman, M. A., H. M. Foley, and J. W. Chamberlain, 1976, Eleven-year variation in polar ozone and stratospheric-ion chemistry, *Science* **192:**555-557.

Schiff, H. I., and J. C. McConnell, 1973, Possible effects of a fleet of supersonic transports on the stratospheric ozone shield, *Rev. Geophys. Space Phys.* **11:**925-934.

Siegenthaler, U., and H. Oeschger, 1978, Predicting future atmospheric carbon dioxide levels, *Science* **199:**388-395.

Solomon, S., P. J. Crutzen, and R. G. Roble, 1982, Photochemical coupling between the thermosphere and the lower atmosphere. 1. Odd nitrogen from 50 to 120 km, *J. Geophys. Res.* **87:**7206-7220.

Starr, W. L., R. A. Craig, M. Loewenstein, and M. E. McGhan, 1980, Measurements of NO, O_3, and temperature at 19.8 km during the total solar eclipse of 26 February 1979, *Geophys. Res. Lett.* **7:**553-555.

Steed, J. M., A. J. Owens, C. Miller, D. L. Filkin, and J. P. Jesson, 1982, Two-dimensional modelling of potential ozone perturbation by chlorofluorocarbons, *Nature,* **295:**308-311.

Stolarski, R. S., and D. M. Butler, 1979, Possible effects of volcanic eruptions on stratospheric minor constituent chemistry, *Pure Appl. Geophys.* **117:**486-497.

Stolarski, R. S., and R. J. Cicerone, 1974, Stratospheric chlorine: A possible sink for ozone, *Canad. J. Chem.* **52:**1610-1615.

Stolarski, R. S., R. J. Cicerone, and A. F. Nagy, 1973, Impact of space shuttle orbiter reentry on mesospheric NO_x, AIAA Paper 73-525.

Sze, N. D., 1977, Anthropogenic CO emissions: Implications for the atmospheric CO-OH-CH$_4$ cycle, *Science* 195:673-675.

Sze, N. D., and H. Rice, 1976, Nitrogen cycle factors contributing to N$_2$O production from fertilizers, *Geophys. Res. Lett.* 3:343-346.

Thorne, R. M., 1977, Energetic radiation belt electron precipitation: A natural depletion mechanism for stratospheric ozone, *Science* 195:287-289.

Turco, R. P., and R. C. Whitten, 1975, Chlorofluoromethanes in the stratosphere and some possible consequences for ozone, *Atmos. Env.* 9:1045-1061.

Turco, R. P., and R. C. Whitten, 1978, A note on the diurnal averaging of aeronomical models, *J. Atmos. Terr. Phys.* 40:13-20.

Turco, R. P., R. C. Whitten, I. G. Poppoff, and L. A. Capone, 1978, SSTs, nitrogen fertilizer and stratospheric ozone, *Nature* 276:805-807.

Turco, R. P., O. B. Toon, C. Park, R. C. Whitten, J. B. Pollack, and P. Noerdlinger, 1981*a,* The Tunguska meteor fall of 1908: Effects on stratospheric ozone, *Science* 214:19-23.

Turco, R. P., O. B. Toon, R. C. Whitten, J. B. Pollack, and P. Hamill, 1981*b,* *Aircraft NO$_x$ Emissions and Stratospheric Ozone Reductions: Another Look,* AIAA Paper 81-0240, Aerospace Sciences Meeting, St. Louis, Missouri, January 12-15.

Turco, R. P., R. C. Whitten, O. B. Toon, E. C. Y. Inn, and P. Hamill, 1981*c,* Stratospheric hydroxyl radical concentrations: New limitations suggested by observations of gaseous and particulate sulfur, *J. Geophys. Res.* 86:1129-1139.

Turco, R. P., O. B. Toon, R. C. Whitten, J. B. Pollack, and P. Noerdlinger, 1982*a,* An analysis of the physical, chemical, optical and historical impacts of the 1908 Tunguska meteor fall, *ICARUS* 50:1-52.

Turco, R. P., O. B. Toon, R. C. Whitten, R. G. Keese, and P. Hamill, 1982*b,* Simulation studies of the physical and chemical processes occurring in the stratospheric clouds of the Mt. St. Helens eruptions of May and June 1980, in *Atmospheric Effects and Potential Climatic Impact of the 1980 Eruption of Mt. St. Helens,* NASA CP-2240, A. Deepak, ed., National Aeronautics and Space Administration, Washington, D.C., 161-189.

Turco, R. P., O. B. Toon, R. C. Whitten, and P. Hamill, 1982*c,* 1-D model simulations of the chemical evolution of the El Chichon eruption cloud, *Am. Geophys. Union Trans.* 63:901.

Turco, R. P., O. B. Toon, R. C. Whitten, R. G. Keese, and P. Hamill, 1983, The 1980 eruptions of Mount St. Helens: Simulation studies of the stratospheric clouds, *J. Geophys. Res.* 88:5299-5319.

U.K. DOE, 1979, *Chlorofluorocarbons and Their Effect on Stratospheric Ozone,* Pollution Paper No. 15, Her Majesty's Stationery Office, London.

Vaughan, G., 1982, Diurnal variation of mesospheric ozone, *Nature* 296:133-135.

Vupputuri, R. K. R., 1979, The structure of the natural stratosphere and the impact of chlorofluoromethane on the ozone layer investigated in a 2-D time dependent model, *Pure Appl. Geophys.* 117:448-485.

Walker, J. C. G., 1977, *Evolution of the Atmosphere,* Macmillan, New York.

Weiss, R. F., 1981, The temporal and spatial distribution of tropospheric nitrous oxide, *J. Geophys. Res.* 86:7185-7195.

Weiss, R. F., and H. Craig, 1976, Production of atmospheric nitrous oxide by combustion, *Geophys. Res. Lett.* 3:751-753.

Whitten, R. C., and I. G. Poppoff, 1971, *Fundamentals of Aeronomy*, John Wiley, New York.

Whitten, R. C., and R. P. Turco, 1974a, Diurnal variations of HO_x and NO_x in the stratosphere, *J. Geophys. Res.* **79**:1302-1304.

Whitten, R. C., and R. P. Turco, 1974b, Perturbations of the stratosphere and mesosphere by aerospace vehicles, *AIAA J.* **12**:1110-1117.

Whitten, R. C., W. J. Borucki, and R. P. Turco, 1975a, Possible ozone depletions following nuclear explosions, *Nature* **257**:38-39.

Whitten, R. C., W. J. Borucki, I. G. Poppoff, and R. P. Turco, 1975b, Preliminary assessment of the potential impact of solid-fueled rocket engines in the stratosphere, *J. Atmos. Sci.* **32**:613-619.

Whitten, R. C., J. Cuzzi, W. J. Borucki, and J. H. Wolfe, 1976, Effect of nearby supernova explosions on atmospheric ozone, *Nature* **263**:398-400.

Whitten, R. C., W. J. Borucki, V. R. Watson, T. Shimazaki, H. T. Woodward, C. A. Riegel, L. A. Capone, and T. Becker, 1977, *The NASA Ames Research Center One- and Two-Dimensional Stratospheric Models, Part II: The Two-Dimensional Model*, Technical Paper 1003, National Aeronautics and Space Administration, Washington, D.C.

Whitten, R. C., W. J. Borucki, L. A. Capone, C. A. Riegel, and R. P. Turco, 1980, Nitrogen fertilizer and stratospheric ozone: Latitudinal effects, *Nature* **283**:191-192.

Whitten, R. C., W. J. Borucki, H. T. Woodward, L. A. Capone, C. A. Riegel, R. P. Turco, I. G. Poppoff, and K. Santhanam, 1981, Implications of smaller concentrations of stratospheric OH: A two-dimensional model study of ozone perturbations, *Atmos. Env.* **15**:1583-1589.

Whitten, R. C., W. J. Borucki, C. Park, L. Pfister, H. T. Woodward, R. P. Turco, L. A. Capone, C. A. Riegel, and T. Kropp, 1982, The Satellite Power System: Assessment of the environmental impact on middle atmosphere composition and on climate *Space Solar Power Rev.* **3**:195-221.

Whitten, R. C., W. J. Borucki, H. T. Woodward, L. A. Capone, and C. A. Riegel, 1983, Revised predictions of the effect on stratospheric ozone of increasing atmospheric N_2O and chlorofluoromethanes: A two-dimensional model study, *Atmos. Env.* **17**:1995-2000.

Widhopf, G. F., L. Glatt, and R. F. Kramer, 1977, Potential ozone column increase resulting from subsonic and supersonic aircraft NO_x emissions, *AIAA J.* **15**:1322-1330.

Wilson, W. J., and P. R. Schwartz, 1981, Diurnal variations of mesospheric ozone using millimeter-wave measurements, *J. Geophys. Res.* **86**:7385-7388.

WMO/NASA, 1982, *The Stratosphere 1981: Theory and Measurements*, Global Ozone Res. Prog. Report No. 11, World Meteorological Organization, Geneva, Switzerland.

Wofsy, S. C., and M. B. McElroy, 1974, HO_x, NO_x and ClO_x: Their role in atmospheric photochemistry, *Can. J. Chem.* **52**:1582-1591.

Wofsy, S. C., J. C. McConnell, and M. B. McElroy, 1972, Atmospheric CH_4, CO, and CO_2, *J. Geophys. Res.* **77**:4477-4493.

Wofsy, S. C., M. B. McElroy, and N. D. Sze, 1975, Freon consumption: Implications for atmospheric ozone *Science* **187**:535-537.

Wofsy, S. C., M. B. McElroy, and Y. L. Yung, 1975, The chemistry of atmospheric bromine, *Geophys. Res. Lett.* **2**:215-218.

Wuebbles, D., and J. S. Chang, 1979, A theoretical study of stratospheric trace species variations during a solar eclipse, *Geophys. Res. Lett.* **6**:179-182.

Wuebbles, D. J., F. M. Luther, and J. E. Penner, 1983, Effect of coupled anthropogenic perturbations on stratospheric ozone, *J. Geophys. Res.* **88**:1444-1456.

Yung, Y. L., M. B. McElroy, and S. C. Wofsy, 1975, Atmospheric halocarbons: A discussion with emphasis on chloroform, *Geophys. Res. Lett.* **2**:397-399.

Yung, Y. L., J. P. Pinto, R. T. Watson, and S. P. Sander, 1980, Atmospheric bromine and ozone perturbations in the lower stratosphere, *J. Atmos. Sci.* **37**:339-353.

Zipf, E. C., and S. S. Prasad, 1980, Production of nitrous oxide in the auroral D and E regions, *Nature* **287**:525-526.

Zipf, E. C., and S. S. Prasad, 1982, A mesospheric source of nitrous oxide, *Nature* **295**:133-135.

Chapter 6

Climatic and Biological Effects

Frederick M. Luther

Lawrence Livermore National Laboratory

The ozone-climate problem has received considerable attention since concern was raised regarding possible threats to stratospheric ozone. Early climatic assessments of reduced ozone focused on the direct solar and longwave effects. Now a number of important feedback mechanisms are recognized as contributing significantly to indirect climatic effects.

Although the focus in this chapter is on the climatic effect of reduced ozone, the discussion must include other trace gases as well. Many of the trace gases that interact photochemically to reduce ozone also have important radiative properties. Examples are chlorofluorocarbons ($CFCl_3$ and CF_2Cl_2), nitrous oxide (N_2O), and methane (CH_4). Other gases, such as CO_2, affect the temperature profile in the atmosphere, which can have an indirect effect on ozone through temperature-dependent reaction rates. The change in ozone, in turn, alters the change in temperature.

The direct radiative effect of gases comes about through absorption of solar radiation and absorption and emission of longwave radiation (also referred to as thermal, terrestrial, or infrared radiation). The spectral distribution of solar and longwave radiation is shown in Figure 6-1. The principal gaseous absorbers of solar radiation are O_2 and O_3 in the stratosphere and H_2O in the troposphere. As discussed in Chapter 2, ozone has absorption bands in the ultraviolet (uv) and visible regions of the solar spectrum. Water vapor absorbs primarily in the near-infrared spectral region.

243

Many gases are radiatively active in the longwave spectral region. Water vapor is a very strong absorber of infrared radiation, effectively blocking longwave radiation emitted at the surface of the earth from being transmitted to space except in the "window" region between 7 and 12 μm. Carbon dioxide is a very effective absorber of longwave radiation at wavelengths near 15 μm. Several gases have strong absorption lines in the 7-12 μm window, among them ozone, chlorofluorocarbons, methane, and nitrous oxide.

Gases that are radiatively active primarily in the longwave spectrum absorb longwave radiation emitted from the earth's surface and emit upward to space and downward to the surface. Since the surface is much warmer than the atmosphere on a global average, the gases absorb more energy than they emit to space. The net result is a reduction of radiation emitted to space by the surface-troposphere system. To maintain a global balance between the net incoming solar radiation and the outgoing longwave radiation, the troposphere and surface warm. This warming is known as the "greenhouse effect," although this is in reality a misnomer.

Ozone and NO_2 differ from the greenhouse gases in that they affect both solar and longwave radiative fluxes. Ramanathan and Dickinson

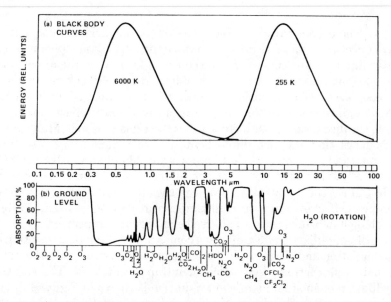

Figure 6-1. The approximate energy distribution of the energy emitted by the sun (6000 K) and the Earth (255 K) shown in relationship to the absorption properties of the Earth's atmosphere. *(From Goody, R. M., 1964, Atmospheric Radiation I: Theoretical Basis, Oxford University Press, London and New York; copyright © 1964 by Oxford University Press.)*

(1979) have shown that an increase in stratospheric ozone could lead to either warming or cooling at the surface, depending on how ozone is redistributed vertically.

In addition to the climatic effect due to a change in ozone, there is the biological effect due to the change in near-uv radiation reaching the earth's surface. This chapter focuses on the climatic and biological effects of a reduction in total ozone.

DIRECT RADIATIVE EFFECTS

Ozone

The temperature structure in the stratosphere is maintained in large part by the absorption of solar radiation by ozone. Figure 6-2 shows the absorption cross-section for ozone. The Hartley band (200-290 nm) and the Huggins band (290-340 nm) modulate the ultraviolet radiation

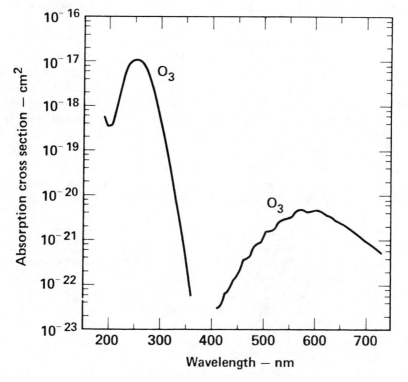

Figure 6-2. The absorption cross-section of ozone.

reaching the troposphere and earth's surface. The Chappuis band (500-700 nm) is much weaker than the other two bands, but it makes an important contribution to the solar heating in the lower stratosphere. The Hartley band provides the major heating at altitudes above about 45 km, the Huggins band dominates between 30 km and 45 km, and the Chappuis band dominates below about 30 km.

The solar heating by ozone leads to an increase of temperature with altitude through the stratosphere. The mixing ratio of ozone is a maximum in the upper stratosphere, which is where the solar heating rate is a maximum. In the upper troposphere, temperature decreases with altitude. The ozone distribution in the stratosphere is primarily responsible for the reversal to temperature increasing with altitude. Consequently, changes in the ozone distribution could affect the altitude of the tropopause and the temperature structure in the stratosphere.

A reduction in stratospheric ozone would lead to a new distribution of solar absorption in the stratosphere and an increase in transmission of solar radiation to the troposphere. Because the Hartley band is so strong, essentially all the solar radiation in this portion of the ultraviolet spectrum is absorbed even with a large reduction of ozone (greater than 50%). With less stratospheric ozone, the radiation would reach lower altitudes before being absorbed. Consequently, there would be a shift in solar heating due to absorption in the Hartley band, but there would be a negligible change in stratospheric transmission. In the weaker Huggins band, a reduction in ozone leads to an increase in stratospheric transmission in the interval from 290-340 nm. Because the Chappuis band is weak, the amount of solar absorption is approximately directly proportional to the amount of ozone. Thus, the change in heating rate due to the Chappuis band is proportional to the change in ozone concentration at that altitude. The change in solar radiation reaching the troposphere occurs in the Huggins and Chappuis bands. Consequently, even though the Chappuis band is weak, it is important in terms of possible climatic effects of changes in stratospheric ozone.

Ozone also absorbs and emits longwave radiation. The strongest vibration-rotation absorption bands are located at 4.75 μm, 9.57 μm, and 14.2 μm. These bands tend to cool the middle and upper stratosphere. In the lower stratosphere these bands can lead to a warming effect at low latitudes. The longwave ozone cooling (or heating) rate depends on the temperature distribution, the ozone distribution, and the background or surface temperature. Because of pressure broadening of absorption lines, the longwave opacity increases with increasing pressure. Consequently, the longwave cooling rates and flux components could change with a redistribution of ozone even though the total column of ozone remained the same.

Table 6-1. Global Mean Stratospheric Radiative Energy Balance.
Net = Absorption − Emission. Units are Wm^{-2}.

Stratos. species	Solar abs.	IR abs.	IR emission	Net
O_3	12.0	7.6	4.1	15.5
$CO_2 + H_2O$	—	19.0	34.9	−15.9
				Total −0.4

Stratospheric effect on the tropospheric energy balance

Stratos. species	Effect on Solar htg of troposphere	IR emission to troposphere	Net
O_3	−7.2	1.7	−5.5
$CO_2 + H_2O$	—	15.7	15.7
			Total 10.2

Source: Data from Ramanathan and Dickerson (1979)

The longwave absorption and emission by ozone provide a greenhouse effect for the troposphere. The 9.57 μm band of ozone is important in this regard since it lies inside the "atmospheric window" between 7 μm and 12 μm, and the distribution of longwave radiation emitted from the surface and lower troposphere is peaked in this region. The other absorption bands of ozone overlap with the strong water vapor absorption bands, which reduce atmospheric transmission at wavelengths on either side of the window region.

The 9.57 μm band of ozone is very effective in absorbing longwave radiation emitted from the earth's surface in the window region. Ozone simultaneously radiates heat upward and downward at its own temperature. This back radiation of energy to the troposphere and surface leads to a warming of the troposphere-surface system.

Ozone plays a very important role in the troposphere-surface energy balance. On an annual and global basis, ozone contributes about 20% of the total longwave flux from the stratosphere to the troposphere (Ramanathan and Dickinson, 1979). The components of the global radiative energy balance are shown in Table 6-1.

The effect of a change in ozone concentration on the downward longwave flux into the troposphere does not involve ozone alone. Changes in stratospheric temperature affect the emission by all other radiatively active constituents in the longwave spectral region. Thus, in the case of a reduction in a stratospheric ozone and no other change in stratospheric composition, there would be less solar heating of the stratosphere and more solar heating of the troposphere (due to greater stratospheric transmission). There would also be a change in the longwave cooling or heating rate, but these changes are small compared to the reduction in the solar heating rate. As the stratosphere cools, the downward longwave emission from the stratosphere to the troposphere by all constituents would be reduced. Consequently, a reduction in stratospheric ozone leads to an increase in the solar flux reaching the troposphere and a reduction in the downward longwave flux into the troposphere (i.e., the solar and longwave effects are opposed in their effect on surface temperature).

Studies using one-dimensional radiative-convective models indicate that a uniform reduction in stratospheric ozone would reduce surface, tropospheric, and stratospheric temperatures. These calculations indicate that a uniform reduction of 20-30% in stratospheric ozone would cool the surface by a few tenths of a kelvin (Reck, 1976; Wang et al., 1976; Ramanathan, Callis, and Boughner, 1976).

Reck (1976) calculated a decrease in surface temperature of 0.6 K for 100% depletion of stratospheric ozone assuming constant absolute humidity and average cloudiness. Manabe and Strickler (1964) also calculated a temperature decrease of less than 1 K for complete removal of the ozone layer. When a low-lying particulate layer was included in the model, Reck (1976) calculated an increase in surface temperature of 0.6 K when the ozone layer was removed. The change in surface temperature also depended on surface albedo.

Ramanathan, Callis, and Boughner (1976) noted that the change in surface temperature is more sensitive to changes in the vertical ozone distribution than to a uniform O_3 reduction. Lowering the altitude of the peak ozone concentration while keeping total ozone constant resulted in an increase in surface temperature.

Changes in stratospheric ozone can have different effects on surface temperature depending on the altitude at which the change in ozone occurs. Changes in ozone concentration in the lower stratosphere and upper troposphere are more effective in causing a surface temperature change than are changes in other altitude regions. Figure 6-3 shows the surface temperature sensitivity to changes in ozone amount at various altitudes for a midlatitude and a tropical atmosphere. The value of the ratio $\Delta T_s / \Delta O_3$ represents the sensitivity of the surface temperature to a change in ozone at that altitude. On a per-molecule basis, ozone changes in the region of 10-12 km are many times more effective than the same change at 20-30 km with respect to surface temperature (Wang, Pinto, and Yung, 1980). These results and the results of Ramanathan and Dickinson (1979) indicate that the change in the radiative flux to the troposphere varies significantly with latitude for the same change in absolute concentration of ozone. The tropics have a larger surface temperature sensitivity in part because the tropics have a smaller column amount of ozone and a higher surface temperature. These results show that a readjustment in the vertical distribution of ozone could have an effect on surface temperature even if the total column amount is unchanged. In the general case, changes in both the ozone concentration profile and total column amount contribute to the climatic effect. Clearly, investigations of the climatic effects of ozone perturbations must consider latitudinal variations of the perturbations.

Ramanathan and Dickinson (1979) considered the latitudinal and seasonal variation of ozone in assessing the climatic effect of an ozone perturbation. Their study also revealed a near balance between the increased solar flux into the troposphere and decreased longwave flux for a uniform reduction in stratospheric ozone. As was the case with the one-dimensional models, a reduction in stratospheric ozone led to cooling at the earth's surface. Their study revealed, however, that the additional solar heating is largely deposited near the surface, whereas the additional

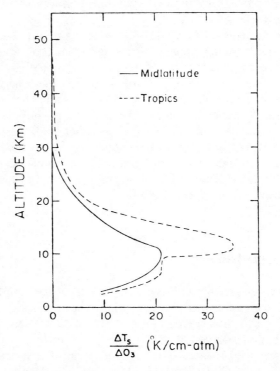

Figure 6-3. Sensitivity of surface temperature change ΔT_s to changes in O_3 concentration for midlatitude and tropical atmospheres. ΔO_3 refers to local O_3 changes (cm-atm) over narrow altitude ranges. *(From Wang, W.-C., J. P. Pinto, and Y. L. Yung, 1980, Climatic effects due to halogenated compounds in the earth's atmosphere, J. Atmos. Sci. **37**; copyright © 1980 by the American Meteorological Society.)*

longwave cooling occurs in the upper troposphere. This difference in heating and cooling in the troposphere would tend to increase the tropospheric lapse rate, which could significantly affect tropospheric dynamics. The change in lapse rate depends on how strongly the surface and upper troposphere are coupled by mixing processes. Since the meridional transport of energy in the atmosphere depends on the meridional temperature gradient, the change in surface temperature also depends on the strength of the coupling between low and high latitudes.

The one-dimensional models usually assume a constant lapse rate, that is, an extremely efficient coupling between the surface and the atmosphere. Consequently, the change in temperature that is computed at the earth's surface applies to all altitudes in the troposphere. The assumption of strong vertical coupling breaks down at high latitudes. Hence, the climatic effects of an ozone perturbation need to be examined with models that explicitly treat dynamical processes in addition to radiative processes.

Although only about 10% of the total column amount of ozone is in the troposphere, changes in tropospheric ozone can lead to climatic effects of about the same magnitude as stratospheric ozone perturbations (Fishman et al., 1979). The longwave opacity of tropospheric ozone is nearly the same as that of stratospheric ozone due to the pressure broadening effect of the 9.6 μm O_3 band. Consequently, uniform percentage changes of tropospheric ozone can have about the same effect on the surface temperature as the same percentage change in stratospheric ozone. For example, Fishman et al. (1979) estimate that halving the tropospheric ozone concentration may cool the surface by 0.5 K, whereas doubling the tropospheric ozone may warm the surface by 0.9 K.

Chlorofluoromethanes

The chlorofluoromethanes $CFCl_3$ and CF_2Cl_2 have strong infrared bands in the spectral region from 8 to 12 μm where the atmosphere is relatively transparent. $CFCl_3$ has strong bands centered at 9.2 and 11.8 μm, and CF_2Cl_2 has bands centered at 8.7, 9.1, and 10.9 μm.

Ramanathan (1975) estimated that increasing the concentrations of both $CFCl_3$ and CF_2Cl_2 to 2 ppbv could raise the surface temperature by 0.9 K. Wang et al. (1976) estimated that the warming would be 0.56 K, and Reck and Fry (1978) estimated 0.60–0.76 K. All these estimates were obtained with one-dimensional radiative-convective models.

Boyer (1979) used a Budyko-Sellers type of energy balance climate model to estimate the warming due to chlorofluoromethanes. For a uniform mixing ratio of 0.7 ppbv of $CFCl_3$ and 1.9 ppbv of CF_2Cl_2, he

Table 6-2. Computed Surface Temperature Change Resulting from Increasing Both $CFCl_3$ and CF_2Cl_2 from O to 2 ppbv

| | Change in Surface Temperature (K) | |
Model	FCA[a]	FCT[b]
Ramanathan (1975)		0.9
Reck and Fry (1978)	0.60–0.76[c]	
Wang et al. (1976)[d]	0.38	0.56
Boyer (1979)		1.4
Wang, Pinto, and Yung (1980)		0.69
Hansen et al. (1981)		0.50
Lacis et al. (1981)		0.65
Ramanathan (1982)	0.56	

[a]Fixed cloud top altitude assumed.
[b]Fixed cloud top temperature assumed.
[c]The smaller value is calculated including particulates; the larger value is without particulates.
[d]Variations of the same radiative model were used by Wang et al. (1976), Wang, Pinto, and Yung (1980), Hansen et al. (1981) and Lacis et al. (1981).

estimated the global warming to be 0.9 K. This sensitivity is higher than the one-dimensional model results for these increases in mixing ratio. The increased sensitivity is attributed to ice-albedo feedback, which leads to larger temperature changes at high latitudes.

More recently, Wang, Pinto, and Yung (1980) estimated the warming to be 0.56 K for an increase of 2 ppbv of $CFCl_3$ and 2 ppbv of CF_2Cl_2. Ramanathan (1982) estimated the warming to be 0.30 K for a 2 ppbv increase in $CFCl_3$ and 0.26 K for a 2 ppb increase in CF_2Cl_2. Hansen et al. (1981) estimated the warming to be 0.50 K for a 2 ppbv increase in $CFCl_3$ and CF_2Cl_2, and Lacis et al. (1981) estimated the warming to be 0.65 K.

The effects of chlorofluoromethanes on surface temperature are summarized in Table 6-2 for the various assessments. In those cases where the effects of $CFCl_3$ and CF_2Cl_2 were evaluated individually, the results are normalized for a 2 ppbv increase in each species concentration.

Methane

Methane has a strong absorption band centered at 7.66 μm. For a doubling of CH_4 from 1.6 ppmv to 3.2 ppmv, Wang et al. (1976) estimated that the surface temperature would increase 0.28–0.40 K, depending on the absorption data used in the calculation. Donner and Ramanathan (1980), Hameed, Cess, and Hogan (1980), and Lacis et al. (1981) all calculated a temperature increase of 0.3 K for a doubling of CH_4 concentration.

Methane concentrations have increased 1-2% per year since 1978 when regular measurements of CH_4 began (Blake et al. 1982). It is not yet clear whether this change represents a long-term trend or just a short-term fluctuation.

Changes in CH_4 concentration affect the global ozone distribution through its reactivity with OH and other trace gases. It is estimated that a doubling of CH_4 would lead to a 2.0% increase in total ozone, with most of the change occurring in the troposphere and lower stratosphere (Wuebbles, Luther, and Penner, 1983).

Nitrous Oxide

Nitrous oxide has strong absorption bands centered at 7.78 μm, 8.56 μm, and 17.0 μm. Wang et al. (1976) calculated a surface warming of 0.44 K for a doubling of N_2O from 280 ppbv to 560 ppbv. Donner and Ramanathan (1980) estimated a surface warming of 0.3 K for a doubling of N_2O. A doubling of N_2O concentrations would lead to increases in stratospheric NO and NO_2, which might decrease total ozone by as much as 15% (WMO, 1981; Whitten et al., 1983; also see Chapter 5). Since measurements began in 1963, N_2O has been increasing at a rate of approximately 0.2% per year (Weiss, 1981).

COUPLED CLIMATE-PHOTOCHEMICAL INTERACTIONS

Ozone is an important constituent in a number of photochemical feedback mechanisms affecting atmospheric composition and climate. In addition to their direct radiative effect, trace gases may interact photochemically with other radiatively important species. For example, surface releases of $CFCl_3$, CF_2Cl_2, N_2O, and CH_4 lead to photochemical products that affect ozone. Increases in N_2O and chlorofluoromethanes have direct radiative effects leading to surface warming, and their dissociation in the stratosphere leads to production of NO_x and ClO_x reactive species that can catalytically destroy ozone (see Chapters 2 and 5). Changes in stratospheric ozone may either cool or warm the surface depending on the vertical distribution of the ozone change, thereby providing either a negative or a positive feedback to the direct N_2O and chlorofluoromethane warming.

In addition to the direct photochemical interaction, the resulting change in stratospheric temperature leads to changes in temperature-dependent chemical reaction rates. The changes in reaction rates in turn alter the species concentration. Thus, there is a coupling between the chemical and the radiative (i.e., temperature) changes.

CFMs, O_3, and Climate

Wang, Pinto, and Yung (1980) investigated the possible effects of the coupling between species concentrations and temperature on the climatic impact of chlorofluoromethanes. The changes in the temperature profile and species concentration profiles were computed at steady state assuming constant CFM release rates at 1973 levels. At steady state the concentrations were 0.8 ppbv for $CFCl_3$ and 2.3 ppbv for CF_2Cl_2, compared to unperturbed values of 0.1 and 0.8 ppbv respectively. The change in surface temperature due to the chlorofluoromethanes alone was a warming of 0.32 K. When the change in ozone was included along with temperature coupling, the increase in surface temperature was only 0.15 K.

Nitrous Oxide, O_3, and Climate

The coupling between chemical and climatic effects of an N_2O perturbation have been investigated by Wang and Sze (1980). N_2O is the most important source of NO_x in the stratosphere. Doubling the tropospheric mixing ratio of N_2O from 320 to 640 ppbv was estimated to increase the surface temperature by 0.44 K when ozone concentrations were held fixed. When ozone was allowed to change, the increased NO_x in the stratosphere resulting from doubling N_2O caused a maximum decrease in ozone concentration of 19% at 38 km and a maximum increase in ozone concentration of 19% at 12 km (Fig. 6-4). The total column amount of ozone was predicted to increase by 2%. The change in species concentrations resulted in a decrease in temperature above 25 km and an increase in temperature below 25 km. With coupling between the chemical and climatic effects, the surface temperature was calculated to increase by 0.62 K when N_2O was doubled. Most of the additional warming compared to a doubling of N_2O alone is due to the temperature increase in the vicinity of the tropopause resulting from increased ozone concentration in that region.

For approximately a doubling of N_2O, Callis, Natarajan, and Boughner (1983) calculated a maximum decrease in ozone concentration of 14% at 37 km. The change in tropospheric ozone concentration ranged from -5% to $+5\%$ depending on the NO_x tropospheric abundance. The total column of ozone was decreased 6.4-6.7% due to a doubling of N_2O. The difference between the results of Wang and Sze and those of Callis et al. is due to advances in our knowledge of the chemical kinetics of stratospheric ozone (see chapter 2 and Whitten et al., 1983).

Carbon Dioxide

Carbon dioxide does not react chemically with ozone. However, by modifying the stratospheric temperature profile, increases in CO_2 concentration can affect ozone through the temperature dependence of chemical rate coefficients. For instance, a doubling of CO_2 leads to an increase in tropospheric temperatures (due to trapping of longwave radiation emitted from the surface and lower troposphere) and a decrease in stratospheric temperatures (due to greater emission to space). The decrease in stratospheric

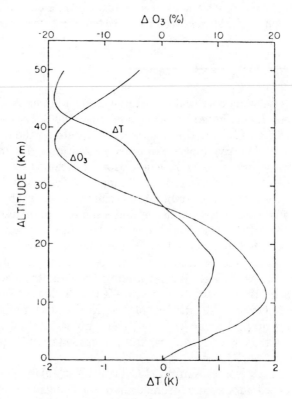

Figure 6-4. The calculated changes in temperature and ozone concentration due to a doubling of present-day N_2O abundance of 320 ppbv. *(From Wang, W.-C., and N. D. Sze, 1980, Coupled effects of atmospheric N_2O and O_3 on the Earth's climate, Nature **286**; copyright © 1980 by Macmillan Journals Ltd.)*

temperatures slows down chemical loss reaction rates, resulting in an increase in stratospheric ozone.

There have been several studies of the coupling between increases in CO_2 concentration and increases in stratospheric ozone (Boughner, 1978; Groves, Mattingley, and Tuck, 1978; Groves and Tuck, 1980; Penner and Luther, 1981; Wuebbles, Luther, and Penner, 1983; Callis, Natarajan, and Boughner, 1983). According to the most recent studies, a doubling of CO_2 might result in an increase in total ozone of 3-4%. The assessments have varied significantly as the understanding of photochemical processes in the stratosphere has evolved. The maximum ozone increase occurs in the upper stratosphere (where the temperature decrease is a maximum) as shown in Figure 6-5. Since there is only a small change in ozone in the lower stratosphere, the CO_2-induced ozone increase has a small effect on surface temperature.

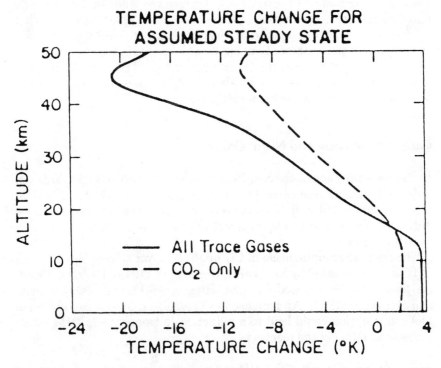

Figure 6-5. The calculated change in temperature when CO_2 is doubled from 300 ppmv, and the temperature change including a doubling of CH_4 and N_2O, 0.7 ppbv of $CFCl_3$ and 1.5 ppbv CF_2Cl_2. *(From Ramanathan, V., 1980, Climatic effects of anthropogenic trace gases, in Interactions of Energy and Climate, W. Bach, J. Pankrath, and J. Williams, eds., Reidel, Dordrecht, Holland; copyright © 1980 by D. Reidel Publishing Company.)*

Water Vapor

Changes in stratospheric water vapor may impact stratospheric ozone by increasing HO_x concentrations. Because of its radiative importance, changes in water vapor would alter stratospheric temperatures and thus also have an indirect impact on stratospheric species concentration. The importance of potential increases in stratospheric water vapor has been studied by Liu et al. (1976) and Luther and Duewer (1978).

The abundance of water vapor in the stratosphere is affected by changes in the temperature of the tropical tropopause (Ellsaesser et al., 1980). Since the saturation vapor pressure depends on temperature, an increase in the tropopause temperature allows more water vapor to enter the stratosphere across the tropical tropopause. The temperature of the tropical tropopause might be altered by increases in the concentrations of trace gases such as $CFCl_3$ and CF_2Cl_2. Using a three-dimensional general circulation model, Dickinson, Liu, and Donahue (1978) estimated that increases in chlorofluoromethane concentrations of 2 ppbv could warm the tropical tropopause by about 2.5 K, which could in turn lead to a 60% increase in water vapor in the lower stratosphere. The increase in water vapor would warm the surface by about 0.5 K (in addition to the surface warming due to the chlorofluoromethanes).

Carbon Monoxide and Nitric Oxide

In the cases cited above, the coupled chemical-climate interaction involved radiatively active trace gases. Other trace gases (such as CO and NO) do not have significant radiative effects of their own, yet they can still have indirect climatic effects via chemical interactions affecting radiatively important gases.

The effect that an increase in CO might have on tropospheric O_3 and CH_4 has been studied by Sze (1977), Logan et al. (1978), Hameed, Pinto, and Stewart (1979), Hameed, Cess, and Hogan (1980), and Callis, Natarajan, and Boughner (1983). An increase in CO resulting from increased fossil fuel consumption could lead to a reduction in tropospheric OH and an increase in HO_2 via the reactions

$$CO + OH \rightarrow CO_2 + H$$

$$H + O_2 \overset{M}{\rightarrow} HO_2 \qquad \qquad (6\text{-}1)$$

Since the reaction of methane with OH ($CH_4 + OH \rightarrow CH_3 + H_2O$) is the only known tropospheric sink for CH_4, a reduction in OH would lead to an increase in methane concentration. A reduction in tropospheric OH could also alter the tropospheric budgets of other gases, including ozone, which are scavenged by OH. Thus, an increase in CO could lead to an increase in tropospheric ozone.

Similarly, an increase in tropospheric NO could lead to an increase in ozone by increasing the rate of the reaction sequence

$$HO_2 + NO \rightarrow NO_2 + OH$$

$$NO_2 + h\nu \rightarrow NO + O \qquad (6\text{-}2)$$

$$O + O_2 \overset{M}{\rightarrow} O_3 .$$

OH is a very important species in terms of chemical and climatic impacts because it is a highly reactive species (see Chapter 2). The abundances of many radiatively important gases are controlled or affected by the amount of OH present. Consequently, any perturbation that alters the OH abundance or distribution could affect the concentrations of many species through coupled chemical processes.

CH$_4$, O$_3$, and Climate

An increase in tropospheric H_2O, which is estimated to occur in response to warming due to increased CO_2, could lead to additional reaction of H_2O with $O(^1D)$ producing more OH. The increased OH would in turn reduce CO and CH_4 and subsequently O_3. The decreases in CH_4 and O_3 would tend to reduce the surface temperature, thus forming a negative climate-chemical feedback response to the initial tropospheric warming.

Another feedback mechanism that couples climatic and chemical processes involves the effect of changes in temperature on the natural emissions of CH_4. A major natural source of CH_4 is anaerobic fermentation of organic material due to microbial action in wetlands, tropical rain forests, and tundra. An increase in surface temperature would lead to increased natural production of CH_4, which would in turn lead to an increase in tropospheric O_3. The radiative effects of CH_4 and O_3 would add to the original warming, thus resulting in a positive feedback mechanism.

The positive and negative feedback mechanisms involving CH_4 have been examined by Hameed and Cess (1983). They found that the negative and positive feedback mechanisms described above had approximately equal compensating effects on the change in global mean temperature. There was less tropospheric O_3 but more tropospheric CH_4 resulting from an increase in surface temperature initially caused by a doubling of CO_2.

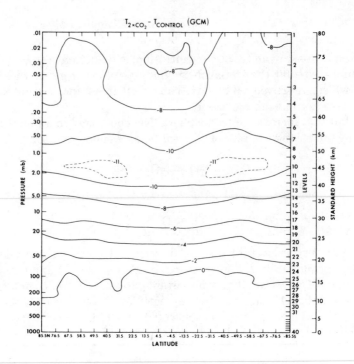

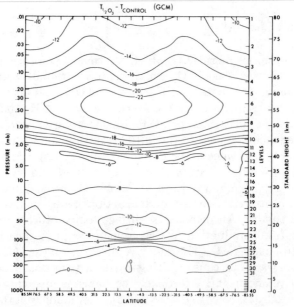

Figure 6-6. Temperature changes due to doubled CO_2 and halved O_3 as simulated by the SKYHI general circulation model. (*From Fels et al., 1980, Stratospheric sensitivity to perturbations in ozone and carbon dioxide: Radiative and dynamical response, J. Atmos. Sci., 37; copyright © 1980 by the American Meteorological Society.*)

Because of uncertainties in tropospheric chemistry and in the parameterization of the effect of changes in temperature on CH_4 production rates, more detailed studies of these feedback mechanisms are still needed.

Atmospheric Transport

Changes in the radiative heating rates and the temperature distribution can lead to changes in atmospheric transport. Schoeberi and Strobel (1978) examined the effect of a uniform decrease in stratospheric O_3 on transport in the middle atmosphere. They computed the change in zonal winds resulting from various reductions in ozone concentrations. The main effect was a weakening of the middle atmosphere jet structures centered at an altitude of about 50-60 km in both hemispheres.

The effect of a uniform reduction in stratospheric ozone was examined by Fels et al. (1980) using a three-dimensional model for the annual-average case. They found that the resulting changes in temperatures varied significantly with latitude (Fig. 6-6), which indicated that these temperature changes might have an effect on atmospheric dynamics. The changes in the dynamics were most important in the tropics and in the mesosphere, with only small changes elsewhere. Fels et al. also found relatively small changes in dynamics due to altered temperature for the case of doubled CO_2. In this case, there was less latitude dependence to the temperature change. Consequently, only small changes in the transports of trace gases are expected to result from these perturbations.

BIOLOGICAL EFFECTS OF CHANGES IN OZONE

Ozone and Ultraviolet Radiation

Ultraviolet radiation is the part of the solar spectrum with wavelengths less than 400 nm. Approximately 9% of the solar flux incident at the top of the atmosphere is in the uv region. Since the energy of a photon is inversely proportional to its wavelength, photons with shorter wavelengths have greater energy. At wavelengths below about 320 nm, photons can produce photochemical damage to the cells of living organisms. Ozone is a very effective absorber of ultraviolet radiation at wavelengths shorter than 300 nm. Consequently, almost no solar radiation with wavelengths shorter than 290 nm reaches the earth's surface. Thus, the biological impact of changes in ozone is due to the amount of uv radiation reaching the ground in the spectral region from 290 to 320 nm, which is called uv-B radiation.

The flux of solar uv radiation reaching the earth's surface on a cloud-less day depends on the amount of attenuation along the optical path due to absorption and scattering of radiation. The amount of stratospheric ozone, trace gases, other air molecules, and aerosol particles all affect the transmission of uv radiation as it passes through the atmosphere. The transmission also depends on the solar zenith angle (the angle of the sun relative to the vertical).

Ozone attenuates uv radiation by direct absorption. The ozone absorp-

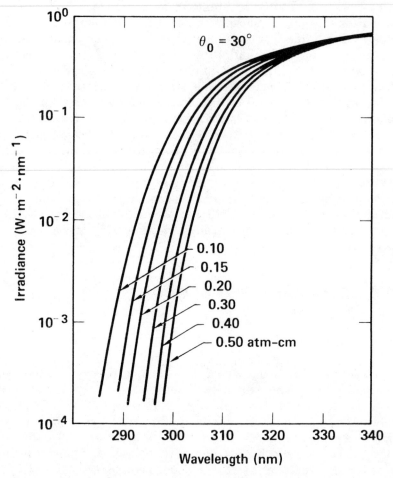

Figure 6-7. Calculated clear-sky solar spectral irradiance for a solar zenith angle of 30° and various amounts of total ozone. *(After Nachtwey, D. S., and R. D. Rundel, 1982, Ozone change: Biological effects, in Stratospheric Ozone and Man, vol. 2, F. A. Bower and R. B. Ward, eds., CRC Press, Boca Raton, Fla.)*

tion coefficient increases sharply as the wavelength decreases from 350 nm to 280 nm (Fig. 6-2). Figure 6-7 shows the clear-sky uv irradiance at the ground for a solar zenith angle of 30° and various column ozone amounts. The irradiance curve represents the product of values of the extraterrestrial irradiance and corresponding values of the transmission curve. When total ozone is reduced, the relative increase in irradiance is greater at the shorter wavelengths. The larger fractional increases occur where the transmission, and consequently the uv irradiance, are small. At wavelengths below 290 nm, the absorption of uv radiation is so effective that there is very little irradiance at these shorter wavelengths even for a 50% reduction in ozone.

Changes in solar zenith angle significantly affect the solar irradiance at all wavelengths, but the fractional changes are larger at the shorter wavelengths where the absorption coefficient is greatest. There have been many theoretical studies of the ultraviolet radiation reaching the earth's surface (Bener, 1970; Green, Sawada, and Shettle, 1974; Halpern, Dave, and Braslau, 1974; Dave and Halpern, 1976; Burt and Luther, 1979; Green, Cross, and Smith, 1980; Gerstl, Zardecki, and Wiser, 1982). These studies focused primarily on clear-sky conditions, and they considered the effect of many factors, including particulates, on the transmission of uv radiation. Other studies considered the influence of clouds on uv radiation (Nack and Green, 1974; Borkowski et al., 1977; Spinhirne and Green, 1978).

The solar irradiance F_λ at wavelength λ that is received at the earth's surface is directly proportional to $\exp[-k_\lambda w_{O_3} \sec \theta_0]$, where k_λ is the absorption coefficient for ozone, w_{O_3} is the total column of ozone, and θ_0 is the solar zenith angle. Then the fractional change in F_λ due to a change in w_{O_3} is

$$\frac{\Delta F_\lambda}{F_\lambda} = -k_\lambda \sec\ \theta_0\ \Delta w_{O_3}. \tag{6-3}$$

Consequently, when k_λ is large, F_λ is small, but the fractional change in F_λ is large for a given reduction in total column ozone. This behavior is illustrated in Figure 6-8, which shows the solar irradiance at the earth's surface as a function of total ozone for different wavelengths for a solar zenith angle of 30°. For example, if total ozone were reduced by 0.03 atm cm (approximately a 10% reduction at middle latitudes), then the irradiance would increase by 106% at a wavelength of 292 nm, but the flux would only increase 10% at 310 nm. However, the flux at 310 nm is about four orders of magnitude larger than the flux at 292 nm in this case.

Total ozone is significantly larger at high latitudes than it is at low latitudes. Also, the daily-average solar zenith angle is larger at high latitudes. Both these factors tend to make F_λ much smaller at high latitudes than at low latitudes. Consequently, there is a strong latitude gradient in the uv flux that is received at the earth's surface and in the biological impact of a reduction in total ozone.

Ultraviolet radiation increases with increasing altitude in the atmosphere. In the lower troposphere, uv radiation is attenuated primarily by scattering by air molecules and particulate material. Moving to higher elevation reduces the amount of air between an observer and the top of the atmosphere. Thus, there is less attenuation of the uv radiation that enters the troposphere. The effect on uv fluxes of moving to higher altitude (lower atmospheric pressure) is shown in Figure 6-9. Since scattering is more effective at the shorter wavelengths, the fractional increase in uv irradiance is greater at the shorter wavelengths as one moves to higher altitude. Moving from sea

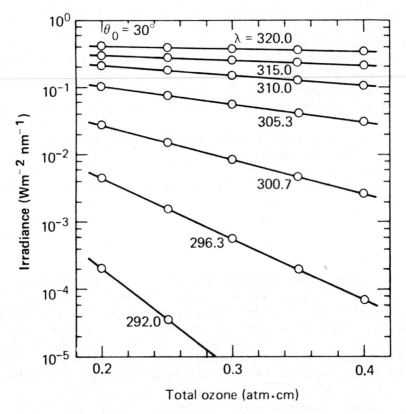

Figure 6-8. Solar irradiance at the Earth's surface for selected wavelengths as a function of total ozone for a solar zenith angle of 30°.

level to an altitude of 2 km results in a 14% increase in irradiance at 300 nm for a solar zenith angle of 30° and total ozone of 0.305 atm cm. This increase is equivalent to the increase that would occur at sea level if ozone were decreased 4%.

The effectiveness of uv radiation in producing a photobiological effect varies significantly with wavelength.

Biologically Effective Ultraviolet Dose

The relative effectiveness of different wavelengths in producing a particular biological effect is called the action spectrum. Each biological effect (e.g., DNA damage, sunburn, plant damage) is characterized by a different action spectrum.

Several action spectra are shown in Figure 6-10. The action spectra are normalized to unity at their maximum value. Curve A is the erythemal (sunburn) action spectrum, which has been used extensively in studies of the biological effects of increased uv radiation on humans (e.g., CIAP,

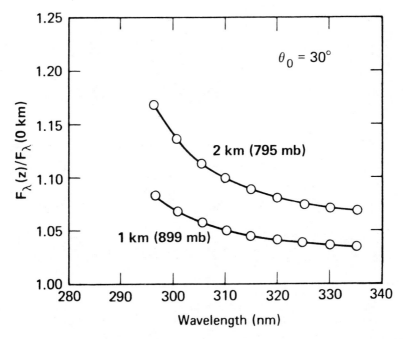

Figure 6-9. The ratio of solar irradiance at 1 and 2 km altitude to that received at sea level for a solar zenith angle of 30°.

1975). Curve B is the response spectrum of the Robertson-Berger Sunburning Ultraviolet Meter, which has been used to measure uv radiation dose rates at a number of geographical locations (National Research Council, 1979, 1982). Curve C is a generalized plant action spectrum developed by Caldwell (1971), and Curve D is the DNA action spectrum developed by Setlow (1974).

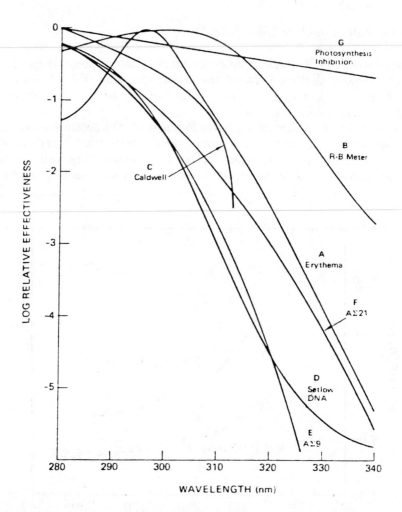

Figure 6-10. Action spectrum weighting functions in current use for biological ultraviolet effects. *(From National Research Council, 1979, Protection Against Depletion of Stratospheric Ozone by Chlorofluorocarbons, National Academy of Sciences, Washington, D.C.)*

In the case where the action spectrum for a particular biological effect is not known, various investigators have used generalized weighting functions based on measured action spectra using comparable biological material. For example, Caldwell (1971) developed a generalized plant action spectrum based on a number of action spectrum studies with plant material, and Setlow (1974) generated a generalized DNA action spectrum based on studies with bacteria and viruses.

The action spectrum for human skin cancer is not known, but it is generally accepted that the spectral sensitivity for skin cancer is similar to either the action spectrum for erythema production or the spectrum for DNA damage (National Research Council, 1975, 1979). The erythemal and DNA action spectra are quite similar in shape, but the DNA action spectrum peaks near 265 nm, whereas the peak in the erythemal action spectrum occurs at 297 nm. The DNA action spectrum was used by the Committee on Biological Effects of Increased Ultraviolet Radiation (National Research Council, 1982) to estimate the potential effect of a reduction in stratospheric ozone on the incidence of skin cancer.

Biological action spectra have been determined by exposing organisms to monochromatic radiation at the same irradiance and scoring the biological response or by supplying the amount of monochromatic flux necessary to elicit a certain threshold response (e.g., erythema). The emphasis has been to determine the structure of the action spectra with as much spectral resolution as possible, but seldom is it possible to determine accurately the tails of the spectra where the response is orders of magnitude less than the peak response. Because the solar irradiance increases rapidly with increasing wavelength within the uv-B region, tails of the action spectra can be quite important in affecting the biologically effective uv dose (Caldwell, 1982).

The biologically effective uv dose is determined by integrating the product of the action spectrum $E(\lambda)$ and the solar irradiance:

$$\text{Effective uv dose} = \int_{\text{uv}} E(\lambda)F(\lambda)\,d\lambda. \qquad (6\text{-}4)$$

The product $E(\lambda)\,F(\lambda)$ is shown in Figure 6-11 for an ozone amount of 0.32 atm cm and for a 50% reduction in total ozone. When ozone is reduced, there is proportionally more uv irradiance at the shorter wavelengths, so the peak in the effective biological irradiance moves to the left (to shorter wavelength). The area under each curve equals the effective uv dose.

The effective uv dose varies gently with latitude and season, reflecting the effect of different solar zenith angles and ozone amounts. The latitude distribution of daily mean erythemal dose for clear-sky conditions is shown in Figure 6-12 for different times of the year. There is a very large latitudinal gradient in the erythemal dose. The annual mean daily erythemal dose varies over two orders of magnitude between the equator and high latitudes (Burt and Luther, 1979). The latitude of the peak daily erythemal dose corresponds closely to the solar declination angle. Other studies of erythemal dose show similar results (Green, Sawada, and Shettle, 1974; Nachtwey and Rundel, 1982).

In relating the effect of a given ozone reduction to a resulting increase in daily biologically effective uv dose, a useful quantity is the radiation amplification factor (RAF). The radiation amplification factor is defined as

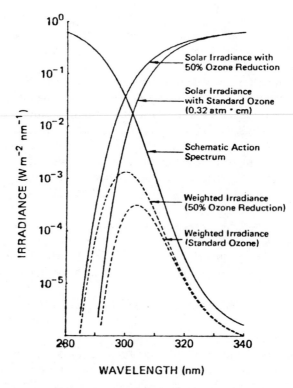

Figure 6-11. Solar irradiance with an ozone thickness of 0.32 atm-cm and with a 50% ozone reduction, superimposed on a typical biological action spectrum. The biologically effective irradiances, represented by the product of the irradiance and the action spectrum, are plotted as dotted curves. *(From National Research Council, 1979, Protection Against Depletion of Stratospheric Ozone by Chlorofluorocarbons, National Academy of Sciences, Washington, D.C.)*

$$\text{RAF} = \frac{\Delta\% \text{ daily effective uv dose}}{\Delta\% \text{ ozone layer thickness}}. \tag{6-5}$$

Values of the RAF have been calculated by Nachtwey and Rundel (1982) for various latitudes and seasons using different action spectra. The annual mean RAF values are shown in Figure 6-13 based on a 1% ozone reduction. For larger reductions in total ozone (say $X\%$), the RAF can be estimated using the values of a 1% ozone reduction:

$$\text{RAF}(X\%) = \text{RAF} (1\%)[1/(1 - 0.01X\%)]. \tag{6-6}$$

The annual mean RAF values for the erythemal action spectrum vary from 1.60 at the equator to 1.75 at 60 N. The RAF values for DNA damage are higher than those for erythema, ranging from 2.21 at the equator to 2.36 at 60 N. The RAF values for the Caldwell action spectrum vary from 1.61 at the equator to 2.17 at 60 N. The RAF values increase with increasing latitude because at higher latitudes both total ozone and the mean solar zenith angle are larger (see Eq. (6-3) to see how this affects irradiance).

According to Eq. (6-3), the percentage increase in uv flux is directly proportional to the ozone absorption coefficient. Shifting the action spectrum to shorter wavelengths results in a greater weighting on wavelengths with larger ozone absorption. Consequently, since the DNA action spectrum is weighted toward shorter wavelengths relative to the erythemal action spectrum, the RAF values are larger for the DNA action spectrum.

Between 0-40 N, the seasonal RAF values are within ±3% of the annual values for the erythemal and DNA action spectra and within ±25% for the Caldwell action spectrum (Nachtwey and Rundel, 1982). The seasonal variation of the RAF values is larger at higher latitudes where the seasonal changes in total ozone are the largest.

Ultraviolet Radiation and Skin Cancer

The damaging effect of uv-B radiation on human beings occurs as the result of prolonged exposure over many years. Sunburn is a short-term effect, which can be very painful and discomforting. More serious, however, is the accumulated effect of repeated exposure to the sun. Most wrinkling of the skin of the face and hands and the warty thickenings of the skin in spots, called solar keratoses, are the result of such exposure. These effects, although undesirable, are not as serious as skin cancer.

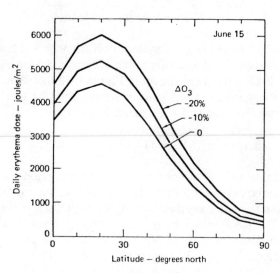

Figure 6-12 *(A)*. Daily total erythemal dose for a horizontal surface as a function of latitude on June 15. *(After Burt, J. E., and F. M. Luther, 1979, Effect of receiver orientation on erythema dose, Photochem. Photobiol.* **29.***)*

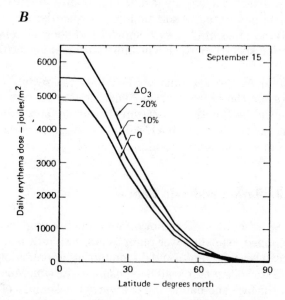

Figure 6-12 *(B)*. Daily total erythemal dose for a horizontal surface as a function of latitude on September 15. *(After Burt, J. E., and F. M. Luther, 1979, Effect of receiver orientation on erythema dose, Photochem. Photobiol.* **29.***)*

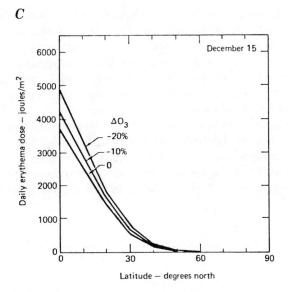

C

Figure 6-12 *(C)*. Daily total erythemal dose for a horizontal surface as a function of latitude on December 15. *(After Burt, J. E., and F. M. Luther, 1979, Effect of receiver orientation on erythema dose, Photochem. Photobiol.* **29.**)

There are three major kinds of skin cancer affecting primarily light-skinned people: basal-cell cancers, squamous-cell cancers, and melanomas. The first two types are the most prevalent, and they are also the most easily and most successfully treated. Basal-cell and squamous-cell cancers (nonmelanomas) usually cause minor disfigurement and seldom lead to death. Because nonmelanoma cancers are slow growing, they will only lead to severe disfigurement or death if they go untreated for a long period of time.

Melanoma, on the other hand, is a serious life-threatening hazard. The occurrence of melanoma is much less frequent than nonmelanoma skin cancers, but there is a significant mortality rate (approximately half of the melanoma patients survive for five or more years). There is a great deal of evidence that uv-B radiation is a cause of nonmelanoma skin cancer. The evidence is not so strong that uv-B radiation is a cause of melanoma skin cancers, but it appears to be a contributing factor.

The pattern of incidence of nonmelanoma skin cancer shows a clear dependence on accumulated uv exposure. The incidence rates increase with decreasing latitude (Fig. 6-14), and nonmelanoma skin cancers are heavily concentrated on the parts of the body most often exposed to sunlight. The face and neck are the most common locations, followed in order by the arms and hands, the trunk, and the legs. Nonmelanoma skin cancers often form in the same tissue systems as sunburn.

There are clear variations among occupational groups. People who work out of doors have a higher rate of incidence, reflecting differences in accumulated dose. Also, the incidence rate of nonmelanoma skin cancers is higher in men than in women, again consistent with occupational differences and rates of exposure.

The incidence rate increases with age, consistent with the dependence on accumulated dose. The incidence is also higher in light-skinned people, particularly those who repeatedly burn with little or no tanning.

It is estimated that exposure to sunlight causes more than 90% of the basal- and squamous-cell skin cancers in the United States (National

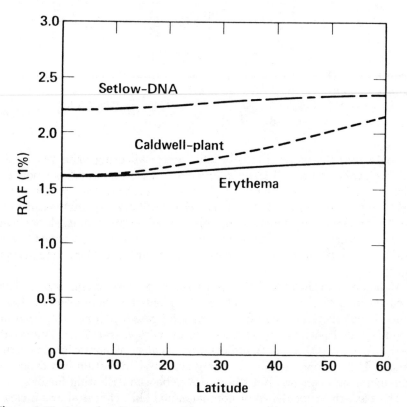

Figure 6-13. Annual radiation amplification factor for a 1% reduction in total ozone calculated for different action spectra. Data from Nachtwey and Rundel (1982).

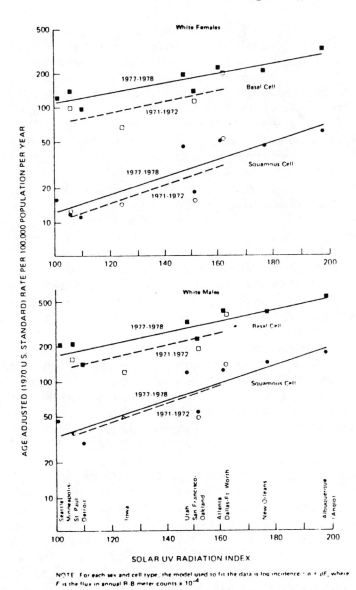

Figure 6-14. Annual age-adjusted incidence rates for basal- and squamous-cell cancers among white females and males for two surveys, 1977–1978 (closed symbols) and 1971–1972 (open symbols). The uv radiation index is the total Robertson–Berger meter counts over a one-year period multiplied by 10⁻⁴. *(From National Research Council, 1982, Causes and Effects of Stratosphere Ozone Reduction, National Academy of Sciences, Washington, D.C.)*

Research Council, 1982). Some nonmelanoma skin cancer is caused by exposure to arsenic, pitch, and x-rays (National Research Council, 1979).

Several studies have estimated the impact of ozone depletion on the incidence rates of basal-cell and squamous-cell skin cancer. One approach is to assume a knowledge of the shape of the dose-response curves from animal data (Rundel and Nachtwey, 1978; DeGruijl and Van der Leun, 1980). The increase in skin cancer as a function of age at a particular location is assumed to follow such a dose-response curve. Using this approach, and assuming that a 1% decrease in ozone causes a 2.3% increase in the uv that would damage DNA, DeGruijl and Van der Leun estimated that a 1% decrease in ozone would cause a 5.5% increase in the incidence rate of nonmelanoma skin cancer.

An EPA (1980) study indicated that for the United States as a whole a 1% decrease in ozone would result in approximately a 4% increase in the incidence rate of nonmelanoma skin cancer. The increase in incidence rate was estimated to be latitude dependent, being larger at low latitudes.

The National Research Council (1982) estimated the effect of a decrease in ozone using the simple model in which the logarithm of the incidence rate of skin cancer is a linear function of annual uv flux. With this model, the estimates of percentage increases in skin cancer incidence that correspond to a given percentage decrease in ozone depend on the magnitude of the annual uv flux. Consequently, the percentage increases in skin cancer incidence are greater at low latitudes. Using the DNA action spectrum to calculate the effect of a reduction in ozone on biologically effective uv flux, it was estimated that a 1% decrease in ozone would lead to a 2 to 5% increase in the incidence rate of basal-cell skin cancer depending on latitude. For squamous-cell skin cancer, the values were approximately double.

The characteristics of the incidence of melanoma are much more complex. They seem to reflect a greater dependence on the dose rate than on the accumulated exposure. Moreover, uv-B seems to be just one of several factors affecting melanoma incidence. The incidence of melanoma has a latitude dependence similar to that for the nonmelanoma skin cancers (National Research Council, 1982; Scotto, Fears, and Fraumeni, 1981). However, there is some deviation from this dependence in certain European countries and in some locations within countries.

As is the case with nonmelanoma skin cancers, most melanomas occur on regularly or occasionally exposed sites, with proportionally fewer on sites that are virtually never exposed. However, the pattern of melanomas over different parts of the body is different from that of nonmelanoma skin cancer. Proportionally more melanomas occur on lightly covered or occasionally uncovered regions of the body and less on the regularly exposed areas. The sites are consistent with those that are exposed to the sun during recreational activity.

The incidence of melanoma has a different distribution among occupational groups. Four studies from very different latitudes showed no significant difference in incidence rates among outdoor and indoor workers of similar status. Socioeconomic status has been shown to be a factor, however. The highest incidence rates occur for individuals with higher socioeconomic status. People in this group tend to spend more time in active outdoor recreational activities, which may indicate that periodic strong exposure to uv-B might contribute more to melanoma than regular moderate exposure. Thus, the incidence rate of melanoma does not show a clear dependence on accumulated exposure.

There is no known direct causal mechanism by which exposure to uv-B radiation leads to melanoma skin cancer. Yet the statistical evidence tends to indicate the uv-B radiation may be involved along with other causal mechanisms. Periodic intense exposure to uv-B may precondition the skin and make it more sensitive to other agents. Because such relationships are only speculative, it is not possible to predict the effect of increased uv-B dose on melanoma incidence.

Plant Responses to Ultraviolet Radiation

Experimental evidence of the effect of uv-B radiation on plants is very limited and of varying quality. Most of the experiments have been conducted during the past 10 years, when concerns were raised about potential threats to stratospheric ozone. The approach has been to expose plants to expected increases in uv-B radiation in a reasonably natural manner. However, problems associated with logistics, technology, and experiment design have made the experiments difficult to conduct for a large number of species and environmental conditions. Consequently, there are limitations on the interpretation and generalization of the results.

In order to control the plant environment effectively, growth chambers have been used so that conditions of temperature, moisture, and uv-B dose can be regulated. Most investigators use fluorescent sunlamps filtered with cellulose acetate or cellulose triacetate to simulate the natural sunlight. The problem with this procedure is that the spectral distribution of the uv irradiance does not agree well with the natural solar uv irradiance. There is too much irradiance at wavelengths less than about 300 nm and too little at longer wavelengths. Consequently, the amount of visible light and uv-A (320-400 nm) radiation available for photoreactivation repair is much less than that available in natural sunlight. Recent evidence strongly suggests that ambient field levels of visible and uv-A radiation can substantially reduce or even negate the damaging effects of uv-B radiation (Sisson and Caldwell, 1976; Teramura, Biggs, and Kossuth, 1980).

The instantaneous dose rate is an important parameter in biological experiments. Nachtwey (1975) showed that the same total dose could have widely differing biological consequences depending on the dose rate. A high dose rate can overwhelm repair processes, leading to greater damage. Many of the earlier experiments conducted in growth chambers used a dose rate much higher than would be found in the natural environment.

Filtered fluorescent sunlamps are also used in field experiments. In this case, the lamps shade the plants, reducing the amount of photosynthetically active visible radiation. The lamps also attract insects, which may affect the plants. Because of the expense and difficulty in irradiating large areas with uv lamps, usually a small sample of plants is used, which increases the statistical uncertainty.

As more has been learned about experiment design and factors that affect plant response, the quality of the experiments has increased. Because the experiments are expensive and time consuming, progress has not been rapid. Therefore, it will be many years before enough data are accumulated to make possible a thorough quantitative assessment of the effect of uv-B radiation on plants. Nevertheless, studies have been made of many plant species and a range of plant response mechanisms (e.g., Bauer et al., 1982) from which conclusions can be made concerning the qualitative effect of uv-B radiation.

More than 100 plant species or varieties have been tested in controlled environmental growth chambers (EPA, 1980; Nachtwey and Rundel, 1982). About one-fifth of these species were very tolerant to increases in uv-B radiation (up to fourfold). About one-fifth were sensitive to present levels of uv-B radiation similar to what is received in the southernmost parts of the United States. The rest of the species had intermediate sensitivities.

Some of the sensitive species and varieties were also tested in the open field as well as in growth chambers. The plants grown in the open field appeared to be much more resistant to uv-B increases. In many cases, there was a fourfold difference in sensitivity. Assuming that the open-field experiments are more representative of actual plant sensitivities, the expected consequences of a reduction in ozone are greatly reduced over what would be projected using the results from the growth chamber experiments. Nevertheless, some agricultural plants, such as many of the row crops (e.g., tomatoes, radishes, cucumber, and mustard), were still found to be sensitive to uv-B increases.

A wide range of sensitivity has been observed in agriculturally important plants (Biggs and Kossuth, 1978). Among the grain crops, soybeans and corn are not very sensitive, but wheat and rice show somewhat greater sensitivity. A wide range of sensitivity has been observed in cultivars within single species (Biggs, Kossuth, and Teramura, 1981; Krizek, 1978), which makes ranking of species by sensitivity impossible. The species also show different sensitivities depending on the type of response being studied (growth, leaf area, seed production, dry weight, protein content, chlorophyll content, etc.). The magnitude of the effect also depends in part on the current level of environmental stress on the plant.

Very few nonagricultural plants have been tested. Those that have been tested have the same range of sensitivity as the crop plants (Caldwell et al., 1975).

Because some plants are more sensitive than others, enhanced uv-B is likely to change the species composition of nonagricultural land ecosystems. The competitive balance among species could be altered, which would change the character of the vegetation.

Effects on Animal and Aquatic Species

Domestic and wild animals do not appear to be very sensitive to enhanced uv-B radiation. The experimental data, although few, do not reveal any large sensitivities. White-faced Hereford cattle, which have very little pigmented material around their eyes, are susceptible to cancer of the eye. Enhanced uv-B would cause an increase in this condition, but only a small number of animals would be affected.

Since the uv-B dose varies greatly with latitude, any large sensitivity to uv-B radiation would be reflected in the incidence rates among animals at different latitudes. There is very limited evidence of latitude-dependent effects of uv-B radiation on animals for effects other than cancer of the eye. Ladds and Entwistle (1977) reported that the incidence of squamous-cell cancers occurring on the ears and nose of sheep was greater in tropical Queensland, Australia than that found in temperate areas in an earlier study (Lloyd, 1961).

Many aquatic species have been tested and found to be sensitive to uv-B radiation. Studies conducted on over 60 aquatic microorganisms, protozoa, algae, and small invertebrates show that most are sensitive to current uv-B radiation levels at the water surface. The attenuation of uv-B radiation in natural waters and the mobility of the organisms apparently play an important role in their survival. The radiation can

Table 6-3. Losses in Aquatic Organisms

Species	Excess Mortality (%)	Increase in uv-b Exposure DNA-Weighted (%)	Data Source
Algae			
(diatoms)	4	38	Van Dyke and Worrest, 1977
Shrimp			Damkaer et al.,
(pandalus species)	50	57	1978
Dungeness crab	50	56	Damkaer et al., 1978
Crab			Damkaer et al.,
(cancer oregonensis)	50	41	1978
Anchovy	80	26	Hunter et al., 1979
	99	50	Hunter et al., 1979
Mackerel	10	26	Hunter et al., 1979
	80	50	Hunter et al., 1979
Freshwater			
microinvertebrates[a]	50	>50	Calkins, 1975

[a]Probable maximum estimate of mortality.

penetrate several meters through clear sea water, so concern about the effect of enhanced uv-B is focused on species that are typically found in rivers, coastal water, and within several meters of the ocean surface.

The sensitivity to enhanced uv-B is highest among the young. Consequently, the change in ozone during the reproductive/development season is more important than the annual-mean value in terms of the impact on aquatic species.

In surface waters, young anchovies normally live near their uv tolerance limit. An increase in uv-B radiation could have a significant effect on this important commercial species. Crab and shrimp larvae have also been shown to be near their uv-B tolerance limit. There has been only limited testing on the uv-B sensitivity of other commercially important aquatic species. The results from several studies are summarized in Table 6-3.

It is difficult to project the impact of an ozone reduction on aquatic species for several reasons. Already mentioned is the limited testing that has been conducted. Another reason is that it is not known whether the species will readjust their usual water depth sufficiently to reduce their uv-B exposure. Such avoidance responses require that the species be able to detect changes in the uv-B irradiance. Some species, particularly the simpler biological micro-organisms, do not have this ability. Additionally, because aquatic communities involve a balance among the various species, the entire community may be affected by reductions in the populations of certain species in the community. Consequently, some species may not be significantly affected by enhanced uv-B directly, but they may be affected indirectly by changes in the population of other species in the aquatic community.

One approach to projecting the aquatic impact of enhanced uv-B is to identify latitudes where species are currently at their uv-B tolerance limit (Nachtwey and Rundel, 1982). Such latitude limits are not easily defined, since many factors affect the latitudinal extent of various species. But based upon experimental results it may be possible to determine the uv-B dose associated with a species' tolerance limit. The latitude at which this dose is received in natural waters then defines the current latitude limit of uv-B tolerance. With an ozone reduction, the latitude limit of uv-B tolerance will move to higher latitude. Presumably, the species population at latitudes equatorward of the new latitude would be adversely affected, and the population at higher latitudes should receive a higher uv-B dose, but the dose would be at or below their tolerance limit. This procedure defines the geographical limits of where the most detrimental effects would be expected. The biological impact within this region would be estimated based on experimental determination of sensitivities and calculated uv-B doses.

SUMMARY

The potential climatic effects of several trace gases have been investigated. In addition to CO_2, other climatically important gases are tropospheric ozone, CH_4, N_2O, $CFCl_3$, and CF_2Cl_2. The change in surface temperature due to potential future increases in the concentrations of these species combined is 50-100% of the temperature change expected from a doubling of CO_2. These gases affect the climate by their direct radiative effects and through coupled chemical-temperature interactions affecting the concentrations of radiatively important species. The largest uncertainties in the climate-chemical assessments come from inadequate understanding of tropospheric chemistry and climate feedback processes.

The effect of a reduction in total ozone on the ultraviolet radiation received at the earth's surface is well understood. The largest uncertainty in calculating the biologically effective radiative dose comes from uncertainty in the action spectra for various biological effects.

The percentage increase in biologically effective uv dose is generally 1.5 to 2.5 times the percentage decrease in total ozone, depending on the latitude, the season, and the action spectrum for the biological effect. Analyses of epidemiological data indicate that the resulting increase in incidence of nonmelanoma skin cancer would be several times the percentage increase in uv dose. Consequently, a reduction in stratospheric ozone could lead to a significantly larger percentage increase in the incidence of nonmelanoma skin cancer. The effect of a reduction in ozone on the incidence of melanoma skin cancer cannot be estimated because there is no clear cause and effect relationship between uv dose and the incidence of melanoma skin cancer.

The effect of an increase in uv radiation on plants, animals, and aquatic species shows a wide range of sensitivities. Because of significant uncertainties in quantifying the sensitivity of each species, it is not yet possible to predict the effect of an ozone reduction on these biological ecosystems.

REFERENCES

Bauer, H., M. M. Caldwell, M. Tevini, and R. C. Worrest, eds., 1982, *Biological Effects of UV-B Radiation,* Geselleschaft für Strahlen- und Umwelfortschung mbH, Munich, FRG.

Bener, P., 1970, Measured and theoretical values of the spectral intensity of ultraviolet zenith radiation and direct solar radiation at 316, 1580, and 2818 m.a.s.l., Air Force Cambridge Research Laboratory, Contract F 61052-67-C-0029, Bedford, Massachusetts.

Biggs, R. H., and S. V. Kossuth, 1978, *Impact of Solar UV-B Radiation on Crop Productivity,* Final Report of UV-B Biological and Climate Effects Research, Terrestrial FY 77, University of Florida, Gainesville.

Biggs, R. H., S. V. Kossuth, and A. H. Teramura, 1981, Response of 19 cultivars of soybeans to ultraviolet-B irradiance, *Physiol. Plant.* 53:19-26.

Blake, D. R., E. W. Mayer, S. C. Tyler, T. Makide, D. C. Montague, and F. S. Rowland, 1982, Global increase in atmospheric methane concentrations between 1978 and 1980, *Geophys. Res. Lett.* 9:477-480.

Borkowski, J., A. T. Chai, T. Mo, and A. E. S. Green, 1977, Cloud effects on middle ultraviolet global radiation, *Acta Geophys. Pol.* 25:287-301.

Boughner, R. E., 1978, The effect of increased carbon dioxide concentrations on stratospheric ozone, *J. Geophys. Res.* 83:1326-1332.

Boyer, G. L., 1979, The "greenhouse" effect of CFMs in a simple energy balance climate model, in *Proceedings of the JOC-GARP Climate Modeling Conference.*

Burt, J. E., and F. M. Luther, 1979, Effect of receiver orientation on erythema dose, *Photochem. Photobiol.* 29:85-91.

Caldwell, M. M., 1971, Solar UV radiation and the growth and development of higher plants, in *Photophysiology,* vol. 6, A. C. Giese, ed., Academic Press, New York, 131-177.

Caldwell, M. M., 1982, Some thoughts on UV action spectra, in *The Role of Solar Ultraviolet Radiation in Marine Ecosystems,* J. Calkins, ed., Plenum, New York, 151-156.

Caldwell, M. M., W. B. Sisson, F. M. Fox, and J. R. Brandle, 1975, Plant growth response to elevated UV irradiance under field and greenhouse conditions, in *Impacts of Climatic Change on the Biosphere, Part 1: Ultraviolet Radiation Effects,* D. S. Nachtwey, M. M. Caldwell, and R. H. Biggs, eds., CIAP Monograph 5, Rept. DOT-TST-75-55, U.S. Department of Transportation, Washington, D.C., 4-253-4-259.

Calkins, J., 1975, Effects of real and simulated solar UV-B in a variety of aquatic microorganisms—possible implication for aquatic ecosystems, in *Impacts of Climatic Change on the Biosphere, Part 1: Ultraviolet Radiation Effects,* D. S. Nachtwey, M. M. Caldwell, and R. H. Biggs, eds., CIAP Monograph 5, Rept. DOT-TST-75-55, U.S. Department of Transportation, Washington, D.C., 5-33-5-69.

Callis, L. B., M. Natarajan, and R. E. Boughner, 1983, On the relationship between the greenhouse effect, atmospheric photochemistry, and species distribution, *J. Geophys. Res.* **88:**1401-1426.

CIAP (Climatic Impact Assessment Program), 1975, *Impacts of Climatic Change on the Biosphere, Part 1: Ultraviolet Radiation Effects,* D. S. Nachtwey, M. M. Caldwell, and R. H. Biggs, eds., CIAP Monography 5, Rept. DOT-TST-75-55, U.S. Department of Transportation, Washington, D.C.

Damkaer, D. M., G. A. Heron, D. B. Deay, and E. F. Prentice, 1978, *Effects of UV-B Radiation on Near-Surface Zooplankton of Puget Sound,* Technical Report, Pacific Marine Environmental Laboratory/NOAA, Seattle, Wash.

Dave, J. V., and P. Halpern, 1976, Effect of changes in ozone amount on the ultraviolet radiation received at sea level of a model atmosphere, *Atmos. Environ.* **10:**547-555.

DeGruijl, F. R., and J. C. Van der Leun, 1980, A dose-response model for skin cancer induction by chronic UV exposure of a human population, *J. Theor. Biol.* **83:**487-504.

Dickinson, R. E., S. C. Liu, and T. M. Donahue, 1978, Effect of chlorofluoromethane infrared radiation on zonal atmospheric temperature, *J. Atmos. Sci.* **35:**2142-2152.

Donner, L., and V. Ramanathan, 1980, Methane and nitrous oxide: Their effects on the terrestrial climate, *J. Atmos. Sci.* **37:**119-124.

Ellsaesser, H. W., J. E. Harries, D. Kley, and R. Penndorf, 1980, Stratospheric H_2O, *Planet. Space Sci.* **28:**827-835.

EPA, 1980, *Results of Research Related to Stratospheric Ozone Protection,* EPA-600/9-80-043, Office of Research and Development, U.S. Environmental Protection Agency, Washington, D.C.

Fels, S. B., J. D. Mahlman, M. D. Schwarzkopf, and R. W. Sinclair, 1980, Stratospheric sensitivity to perturbations in ozone and carbon dioxide: Radiative and dynamical response, *J. Atmos. Sci.* **37:**2265-2297.

Fishman, J., V. Ramanathan, P. J. Crutzen, and S. C. Liu, 1979, Tropospheric ozone and climate, *Nature* **282:**818-820.

Gerstl, S. A. W., A. Zardecki, and H. L. Wiser, 1981, Biologically damaging radiation amplified by ozone depletions, *Nature* **294:**352-354.

Goody, R. M., 1964, *Atmospheric Radiation I: Theoretical Basis,* Oxford University Press, London and New York.

Green, A. E. S., K. R. Cross, and L. A. Smith, 1980, Improved analytic characterization of ultraviolet skylight, *Photochem. Photobiol.* 31:59-65.

Green, A. E. S., T. Sawada, and E. P. Shettle, 1974, The middle ultraviolet reaching the ground, *Photochem. Photobiol.* 19:251-262.

Groves, K. S., and A. F. Tuck, 1980, Stratospheric O_3-CO_2 coupling in a photochemical-radiative column model, *R. Met. Soc. Q. J.* 106:125-157.

Groves, K. S., S. R. Mattingly, and A. F. Tuck, 1978, Increased atmospheric carbon dioxide and stratospheric ozone, *Nature* 273:711-715.

Halpern, P., J. V. Dave, and N. Braslau, 1974, Sea-level solar radiation in the biologically active spectrum, *Science* 186:1204-1208.

Hameed, S., and R. D. Cess, 1983, Impact of a global warming on biospheric sources of methane and its climatic impact, *Tellus* 35B:1-7.

Hameed, S., R. D. Cess, and J. Hogan, 1980, Response of the global climate to changes in atmospheric chemical composition due to fossil fuel burning, *J. Geophys. Res.* 85:7537-7545.

Hameed, S., J. P. Pinto, and R. W. Stewart, 1979, Sensitivity of the predicted CO-OH-CH_4 perturbation to tropospheric NO_x concentrations, *J. Geophys. Res.* 84:763-768.

Hansen, J., D. Johnson, A. Lacis, S. Lebedeff, P. Lee, D. Rind, and G. Russell, 1981, Climate impact of increasing atmospheric carbon dioxide, *Science* 213:957-966.

Hunter, J. R., J. H. Taylor, and H. G. Moser, 1979, Effect of ultraviolet irradiation on eggs and larvae of the northern anchovy, *Engraulis mordax,* and the Pacific mackerel, *Scomber japonicus,* during the embryonic stage, *Photochem. Photobiol.* 29:325-338.

Krizek, D. T., 1978, *Differential Sensitivity of Two Cultivars of Cucumber* (Cucumis sativus L.) *to increased UV-B Irradiance. I. Dose-Response Studies,* Final Report of UV-B Biological and Climate Effects Research, Terrestrial FY 77, Beltsville Agricultural Research Center, Beltsville, Md.

Lacis, A., J. Hansen, P. Lee, T. Mitchell, and S. Lebedeff, 1981, Greenhouse effect of trace gases, 1970-1980, *Geophys. Res. Lett.* 8:1035-1038.

Ladds, P. W., and K. W. Entwistle, 1977, Observations on squamous cell carcinomas of sheep in Queensland, Australia, *Br. J. Cancer* 35:110-114.

Liu, S. C., T. M. Donahue, R. J. Cicerone, and W. L. Chameides, 1976, Effect of water vapor on the destruction of ozone in the stratosphere perturbed by Cl_x or NO_2 pollutants, *J. Geophys. Res.* 81:3111-3118.

Lloyd, L. C., 1961, Epithelial tumors of the skin of sheep, *Br. J. Cancer* 15:780.

Logan, J. A., J. J. Prather, S. C. Wofsy, and M. B. McElroy, 1978, Atmospheric chemistry: Response to human influences, *R. Soc. London Philos. Trans.* 290:187-234.

Luther, F. M., and W. H. Duewer, 1978, Effect of changes in stratospheric water vapor on ozone reduction estimates, *J. Geophys. Res.* 83:2395-2402.

Manabe, S., and R. F. Strickler, 1964, Thermal equilibrium of the atmosphere with a convection adjustment, *J. Atmos. Sci.* 21:361-385.

Nack, L. M., and A. E. S. Green, 1974, Influence of clouds, haze, and smog on the middle ultraviolet reaching the ground, *Appl. Opt.* 13:2405-2415.

Nachtwey, D. S., 1975, Dose rate effects of the UV-B inactivation of chlamydomonas and implications for survival in nature, in *Impacts of Climatic Change on the Biosphere, Part 1, Ultraviolet Radiation Effects,* D. S. Nachtwey, M. M. Caldwell, and R. H. Biggs, eds., CIAP Monograph 5, Rept. DOT-TST-75-55, U.S. Dept. of Transportation, Washington, D.C., 3-105-3-119.

Nachtwey, D. S., and R. D. Rundel, 1982, Ozone change: Biological effects, in *Stratospheric Ozone and Man,* vol. 2, F. A. Bower and R. B. Ward, eds., CRC Press, Boca Raton, Fla., 81-121.

National Research Council, 1975, *Environmental Impact of Stratospheric Flight: Biological and Climatic Effects of Aircraft Emissions in the Stratosphere,* National Academy of Sciences, Washington, D.C.

National Research Council, 1979, *Protection Against Depletion of Stratospheric Ozone by Chlorofluorocarbons,* National Academy of Sciences, Washington, D.C.

National Research Council, 1982, *Causes and Effects of Stratospheric Ozone Reduction,* National Academy of Sciences, Washington, D.C.

Penner, J. E., and F. M. Luther, 1981, Effect of temperature feedback and hydrostatic adjustment in a stratospheric model, *J. Atmos. Sci.* 38:446-453.

Ramanathan, V., 1975, Greenhouse effect due to chlorofluorocarbons: Climatic implications, *Science* 190:50-52.

Ramanathan, V., 1980, Climatic effects of anthropogenic trace gases, in *Interactions of Energy and Climate,* W. Bach, J. Pankrath, and J. Williams, eds., Reidel, Dordrecht, Holland, 269-280.

Ramanathan, V., 1982, Commentary on "Climate effects of minor atmospheric constituents," in *Carbon Dioxide Review,* Oxford University Press, New York, 278-283.

Ramanathan, V., and R. E. Dickinson, 1979, The role of stratospheric ozone in the zonal and seasonal radiative energy balance of the earth-troposphere system, *J. Atmos. Sci.* 36:1084-1104.

Ramanathan, V., L. B. Callis, and R. E. Boughner, 1976, Sensitivity of surface temperature and atmospheric temperature to perturbations in the stratospheric concentrations of ozone and nitrogen dioxide, *J. Atmos. Sci.* 33:1092-1112.

Reck, R. A., 1976, Stratospheric ozone effects on temperature, *Science* 192:557-559.

Reck, R. A., and D. L. Fry, 1978, The direct effects of CFMs on the atmospheric and surface temperatures, *Atmos. Environ.* 12:2501-2503.

Rundel, R. D., and D. S. Nachtwey, 1978, Skin cancer and ultraviolet radiation, *Photochem. Photobiol.* 28:345-356.

Schoeberl, M. R., and D. F. Strobel, 1978, The response of the zonally averaged circulation to stratospheric ozone reductions, *J. Atmos. Sci.* 35:1751-1757.

Scotto, J., T. R. Fears, and J. F. Fraumeni, Jr., 1981, *Incidence of Non-melanoma Skin Cancer in the United States,* DHHS Publication No. (NIH) 82-2433, National Cancer Institute, Bethesda, Md.

Setlow, R. B., 1974, The wavelengths in sunlight effective in producing skin cancer: A theoretical analysis, *Natl. Acad. Sci. (USA) Proc.* 71:3363-3366.

Sisson, W. B., and M. M. Caldwell, 1976, Photosynthesis, dark respiration, and growth of Rumex patientia L. exposed to ultraviolet irradiance (288-315 nm) simulating a reduced atmospheric ozone column, *Plant Physiol.* 58:563-568.

Spinhirne, J. D., and A. E. S. Green, 1978, Calculation of the relative influence of cloud layers on received ultraviolet and integrated solar radiation, *Atmos. Environ.* **12**:2449-2454.

Sze, N. D., 1977, Anthropogenic CO emission: Implications for atmospheric CO-OH-CH$_4$ cycle, *Science* **195**:673-675.

Teramura, A. H., R. H. Biggs, and S. V. Kossuth, 1980, Effects of ultraviolet-B irradiance on soybean. II. Interaction of ultraviolet-B and photosynthetically active radiation on net photosynthesis, dark respiration, and transpiration, *Plant Physiol.* **65**:483-488.

Van Dyke, H., and R. C. Worrest, 1977, *Assessment of the Impact of Increased Solar Ultraviolet Radiation upon Marine Ecosystems,* NAS 9-14860, Mod. 78, NASA Lyndon B. Johnson Space Center, Houston, Tex.

Wang, W.-C., and N. D. Sze, 1980, Coupled effects of atmospheric N$_2$O and O$_3$ on the Earth's climate, *Nature* **286**:589-590.

Wang, W.-C., J. P. Pinto, and Y. L. Yung, 1980, Climatic effects due to halogenated compounds in the earth's atmosphere, *J. Atmos. Sci.* **37**:333-338.

Wang, W.-C., Y. L. Yung, A. A. Lacis, T. Mo, and J. E. Hansen, 1976, Greenhouse effects due to man-made perturbations of trace gases, *Science* **194**:685-690.

Weiss, R. W., 1981, The temporal and spatial distribution of tropospheric nitrous oxide, *J. Geophys. Res.* **86**:7185-7195.

Whitten, R. C., W. J. Borucki, H. T. Woodward, L. A. Capone, and C. A. Riegal, 1983, Revised prediction of the effect on stratospheric ozone of increased atmospheric N$_2$O and chlorofluoromethanes: A two-dimensional model study, *Atmos. Environ.* **17**:1995-2000.

WMO, 1981, *The Stratosphere 1981: Theory and Measurements,* WMO Global Ozone Research and Monitoring Project Report No. 11, World Meteorological Organization, Geneva, Switzerland.

Wuebbles, D. J., F. M. Luther, and J. E. Penner, 1983, Effect of coupled anthropogenic perturbations on stratospheric ozone, *J. Geophys. Res.* **88**:1444-1456.

Index